DeWALT®

ELECTRICAL

PROFESSIONAL REFERENCE

2011 CODE

Paul Rosenberg

Published by:

DELMAR
CENGAGE Learning™

Australia • Brazil • Japan • Korea • Mexico • Singapore • Spain • United Kingdom • United States

dewalt.com/guides

OTHER TITLES AVAILABLE

This Book Belongs To:

Name: ________________________________

Company: ________________________________

Title: ________________________________

Department: ________________________________

Company Address: ________________________________

____

____

Company Phone: ________________________________

Home Phone: ________________________________

DELMAR
CENGAGE Learning

DeWALT Electrical Professional Reference: 2011 Code
Paul Rosenberg

Vice President, Technology and Trades Professional Business Unit: Gregory L. Clayton

Director of Building Trades: Taryn Zlatin McKenzie

Product Development Manager: Robert Person

Development Editor: Nobina Chakraborti

Content Project Management: PreMediaGlobal

Director of Marketing: Beth A. Lutz

Marketing Manager: Marissa Maiella

Production Director: Carolyn Miller

Production Manager: Andrew Crouth

Manufacturing Buyer: Denise Powers

Art Director: Benjamin Gleeksman

Production Service/Compositor: PreMediaGlobal

Library of Congress Control Number: 2010939204

ISBN-13: 978-1-111-54514-7
ISBN-10: 1-111-54514-6

Delmar
5 Maxwell Drive, Clifton Park, NY 12065-2919, USA

Cengage Learning is a leading provider of customized learning solutions with office locations around the globe, including Singapore, the United Kingdom, Australia, Mexico, Brazil and Japan. Locate your local office at: **international.cengage.com/region**

Cengage Learning products are represented in Canada by Nelson Education, Ltd.

Visit us at www.InformationDestination.com

For more learning solutions, please visit our corporate website at **www.cengage.com**

Notice to the Reader
Cengage Learning and DeWALT do not warrant or guarantee any of the products described herein or perform any independent analysis in connection with any of the product information contained herein. Cengage Learning and DeWALT do not assume, and expressly disclaim, any obligation to obtain and include information other than that provided to it by the manufacturer. The reader is expressly warned to consider and adopt all safety precautions that might be indicated by the activities described herein and to avoid all potential hazards. By following the instructions contained herein, the reader willingly assumes all risks in connection with such instructions. Cengage Learning and DeWALT make no representations or warranties of any kind, including but not limited to, the warranties of fitness for particular purpose or merchantability, nor are any such representations implied with respect to the material set forth herein, and Cengage Learning and DeWALT take no responsibility with respect to such material. Cengage Learning and DeWALT shall not be liable for any special, consequential, or exemplary damages resulting, in whole or part, from the readers' use of, or reliance upon, this material.

Printed in China

1 2 3 4 5 6 7 14 13 12 11 10

CONTENTS

CHAPTER 4 – *Receptacles, Switches, Interior Wiring and Lighting* 4-1

Preface

Many years ago I was asked to create a compendium of tables, charts, graphs, diagrams and terminology relating to the field of electricity. Electrical Reference was my first endeavor to create a pocket guide for the electrical marketplace that would serve as an all encompassing publication in a new user-friendly format. I ended up utilizing a design and typeface that holds the most information per page while at the same time making a particular topic quick to reference.

Information covered in this manual is necessary for anyone in the field to have in one's possession at all times. Naturally, a topic may have been overlooked or not discussed in depth to suit all tradespeople. I will constantly monitor and update this book on a regular basis to not only include requested additional material, but to add new material from the ever growing amount of high technology as it develops.

Best wishes,
Paul Rosenberg

A Note To The Reader

Electrical Professional Reference is not a substitute for the National Electric Code®. National Electric Code® and NEC® are registered trademarks of the National Fire Protection Association, Inc. Quincy, MA.

CHAPTER 1
Circuits, Formulas and Voltage Drop Calculations

OHM'S LAW/POWER FORMULAS

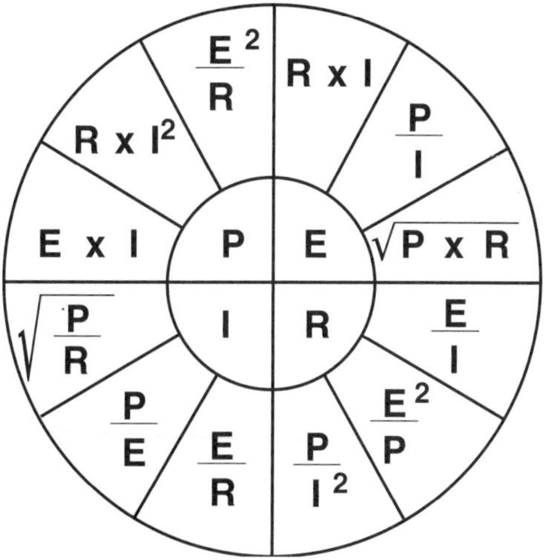

P = Power = Watts

R = Resistance = Ohms

I = Current = Amperes

E = Force = Volts

OHM'S LAW DIAGRAM AND FORMULAS

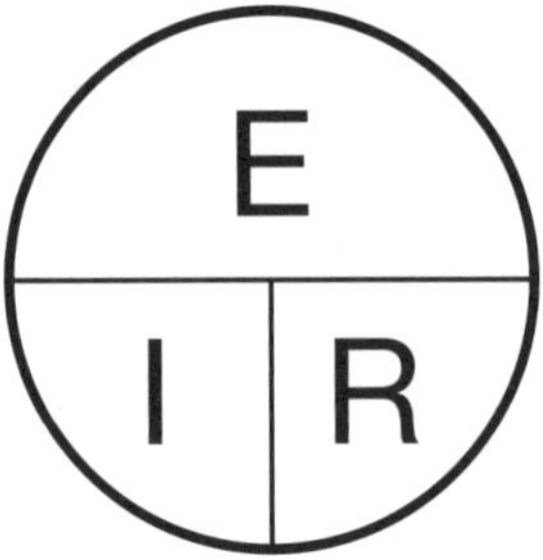

$E = I \times R$ Voltage = Current × Resistance
$I = E \div R$ Current = Voltage ÷ Resistance
$R = E \div I$ Resistance = Voltage ÷ Current

POWER DIAGRAM AND FORMULAS

$P = I \times E$ Power = Current × Voltage
$E = P \div I$ Voltage = Power ÷ Current
$I = P \div E$ Current = Power ÷ Voltage

OHM'S LAW AND IMPEDANCE

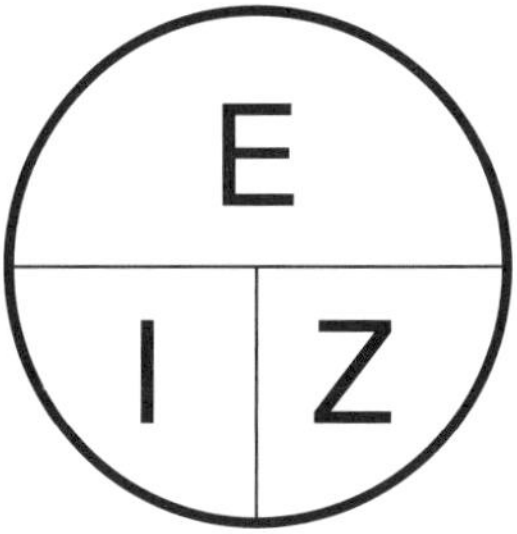

E = VOLTAGE (IN VOLTS)
I = CURRENT (IN AMPS)
Z = IMPEDANCE (IN OHMS)

$E = I \times Z$ $\quad\quad I = \frac{E}{Z}$ $\quad\quad Z = \frac{E}{I}$

Ohm's law and the power formula are limited to circuits in which electrical resistance is the only significant opposition to current flow including all DC circuits and AC circuits that do not contain a significant amount of inductance and/or capacitance. AC circuits that include inductance are any circuits that include a coil as the load such as motors, transformers, and solenoids. AC circuits that include capacitance are any circuits that include a capacitor(s).

In DC and AC circuits that do not contain a significant amount of inductance and/or capacitance, the opposition to current flow is resistance (R). In circuits that contain inductance (X_L) or capacitance (X_C), the opposition to the flow of current is reactance (X). In circuits that contain resistance (R) and reactance (X), the combined opposition to the flow of current is impedance (Z). Resistance and Impedance are both measured in Ohms.

Ohm's law is used in circuits that contain impedance, however, and Z is substituted for R in the formula. Z represents the total resistive force (resistance and reactance) opposing current flow.

AC/DC POWER FORMULAS

To Find	Direct Current	Alternating Current		
		Single Phase	Two-Phase Four-Wire*	Three Phase
Amps when horsepower is known	$\frac{HP \times 746}{E \times E_{FF}}$	$\frac{HP \times 746}{E \times E_{FF} \times PF}$	$\frac{HP \times 746}{2 \times E \times E_{FF} \times PF}$	$\frac{HP \times 746}{1.73 \times E \times E_{FF} \times PF}$
Amps when kilowatts are known	$\frac{kW \times 1000}{E}$	$\frac{kW \times 1000}{E \times PF}$	$\frac{kW \times 1000}{2 \times E \times PF}$	$\frac{kW \times 1000}{1.73 \times E \times PF}$
Amps when kVA is known	—	$\frac{kVA \times 1000}{E}$	$\frac{kVA \times 1000}{2 \times E}$	$\frac{kVA \times 1000}{1.73 \times E}$
Kilowatts	$\frac{I \times E}{1000}$	$\frac{I \times E \times PF}{1000}$	$\frac{I \times E \times 2 \times PF}{1000}$	$\frac{I \times E \times 1.73 \times PF}{1000}$
Kilovolt-Amps	—	$\frac{I \times E}{1000}$	$\frac{I \times E \times 2}{1000}$	$\frac{I \times E \times 1.73}{1000}$
Horsepower	$\frac{I \times E \times E_{FF}}{746}$	$\frac{I \times E \times E_{FF} \times PF}{746}$	$\frac{I \times E \times 2 \times E_{FF} \times PF}{746}$	$\frac{I \times E \times 1.73 \times E_{FF} \times PF}{746}$

I = Amps; E = Volts; E_{FF} = Efficiency; PF = Power Factor; kW = Kilowatts; kVA = Kilovolt Amps; HP = Horsepower
*For three wire, two phase circuits the current in the common conductor is 1.41 times the current in either of the other two conductors.

Power Factor = cos (Phase Angle)
Power Factor = True Power / Apparent Power
Power Factor = Watts / Volts × Amps
Power Factor = Resistance (in Ohms) / Impedance (in Ohms)
Efficiency = Output / Input

THREE-PHASE AC CIRCUITS AND THE UTILIZATION OF POWER

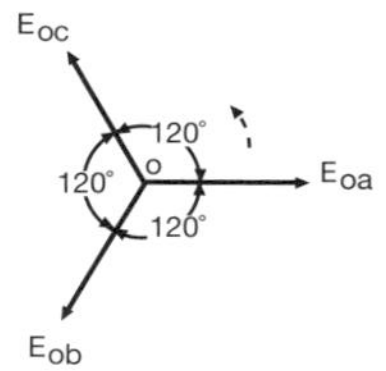

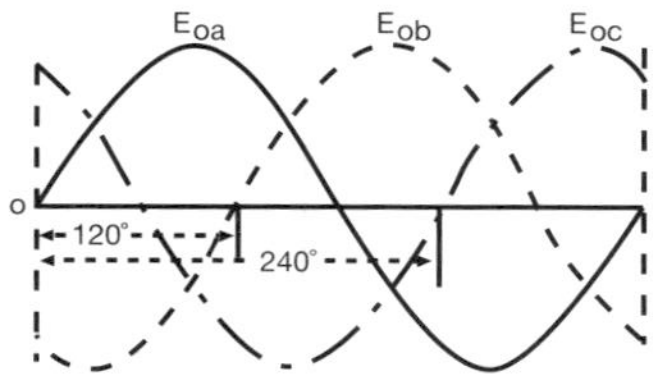

Sine waves are actually an oscillograph trace taken at any point in a three-phase system. (Each voltage or current wave actually comes from a separate wire but are shown for comparison on common base). There are 120° between each voltage. At any instant the algebraic sum (measured up and down from centerline) of these three voltages is zero. When one voltage is zero, the other two are 86.6% maximum and have opposite signs.

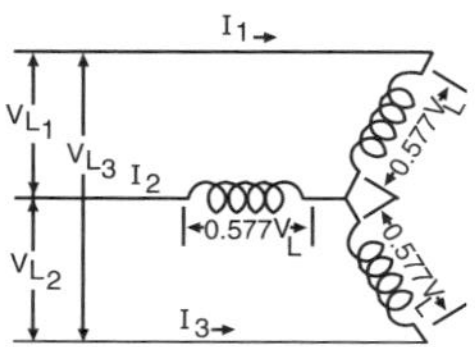

WYE CONNECTION

Consider the three windings as primary of transformer. Current in all windings equals line current, but volts across windings = 0.577 × line volts.

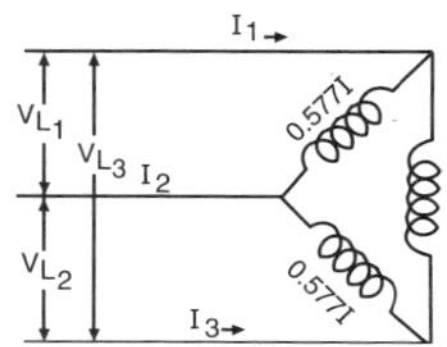

DELTA CONNECTION

Winding voltages equal line voltages, but currents split up so 0.577 I_{line} flows through windings.

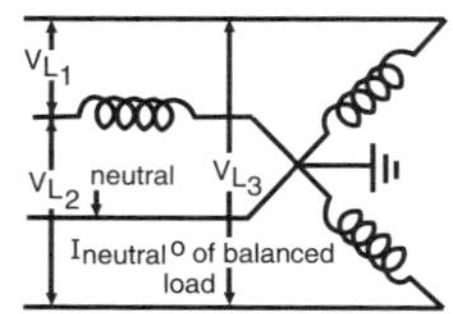

FOUR-WIRE SYSTEM

Most popular secondary distribution setup. V_1 is usually 208 V which feeds small power loads. Lighting loads at 120 V tap from any line to neutral.

MULTI-WIRE SYSTEM VOLTAGES

Wye-Connected

120/208 V, 3φ, 4-wire wye

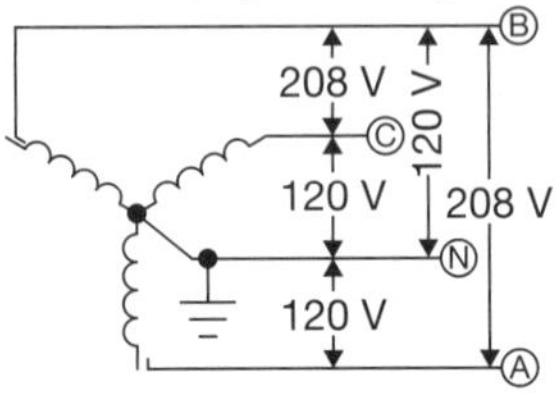

A to B = 208 V
B to C = 208 V
A to N = 120 V
B to N = 120 V
C to N = 120 V

277/480 V, 3φ, 4-wire wye

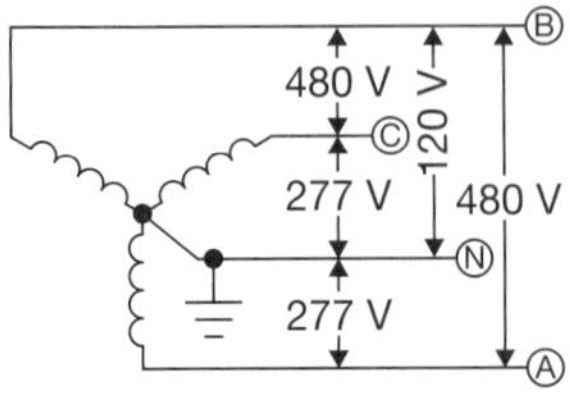

A to B = 480 V
B to C = 480 V
A to N = 277 V
B to N = 120 V
C to N = 277 V

Delta-Connected

120/240 V, 3ϕ, 4-wire delta

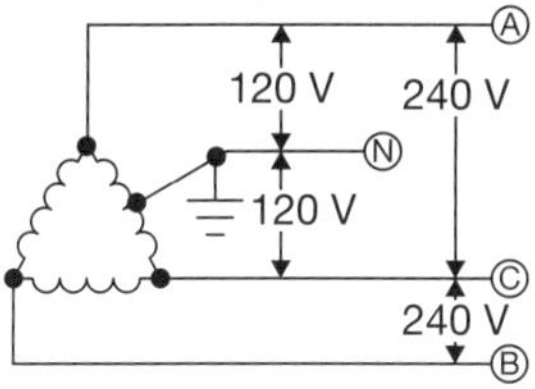

A to B = 240 V
B to C = 240 V
A to N = 120 V
B to N = 120 V
C to N = Not used

480 V, 3ϕ, 3-wire delta

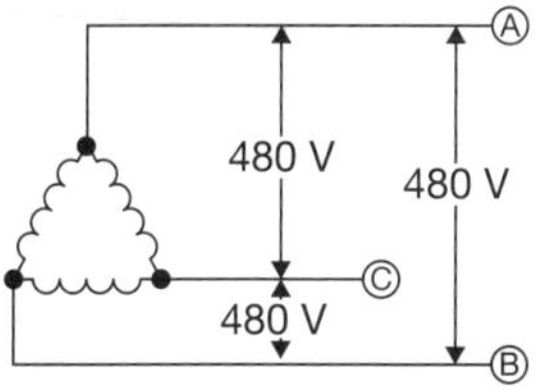

A to B = 480 V
B to C = 480 V
A to C = 480 V

FORMULAS FOR SINE WAVES

Frequency	Period	RMS Voltage
$F = \frac{1}{T}$ where F = Frequency (in hertz) T = Period (in seconds) 1 = Constant	$T = \frac{1}{F}$ where T = Period (in seconds) F = Frequency (in hertz) 1 = Constant	$V_{RMS} = V_{MAX} \times .707$ where V_{RMS} = RMS voltage V_{MAX} = Peak voltage .707 = Constant

Average Voltage	Peak-to-Peak Voltage
$V_{AVG} = V_{MAX} \times .637$ where V_{AVG} = Average voltage V_{MAX} = Peak voltage .637 = Constant	$V_{P\text{-}P} = V_{MAX} \times 2$ where $V_{P\text{-}P}$ = Peak-to-peak voltage V_{MAX} = Peak voltage 2 = Constant

CALCULATING PEAK, AVERAGE AND ROOT-MEAN-SQUARE (RMS) VOLTAGES

Effective (RMS) voltage = 0.707 × Peak voltage
Effective (RMS) voltage = 1.11 × Average voltage
Average voltage = 0.637 × Peak voltage
Average voltage = 0.9 × Effective (RMS) voltage
Peak voltage = 1.414 × Effective (RMS) voltage
Peak voltage = 1.57 × Average voltage
Peak-to-Peak voltage = 2 × Peak voltage
Peak-to-Peak voltage = 2.828 × Effective (RMS) voltage

KIRCHHOFF'S LAWS

First Law of Current

The sum of the currents arriving at any point in a circuit must equal the sum of the currents leaving that point.

Second Law of Voltage

The total voltage applied to any closed circuit path is always equal to the sum of the voltage drops in that path.

POWER FACTOR

An AC electrical system carries two types of power: (1) true power, watts, that pulls the load (note: mechanical load reflects back into an AC system as resistance.) and (2) reactive power, vars, that generates magnetism within inductive equipment. The vector sum of these two will give actual volt-amperes flowing in the circuit (see diagram below). Power factor is the cosine of the angle between true power and volt-amperes.

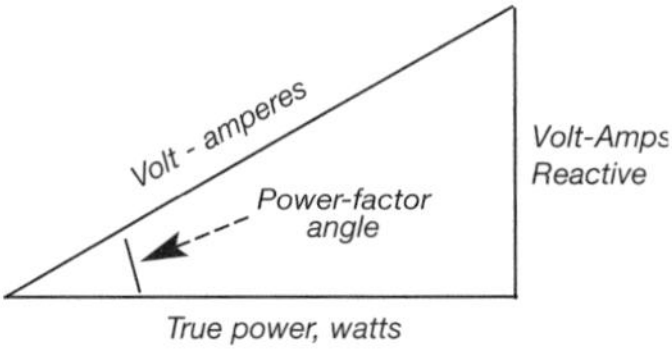

TYPES OF POWER

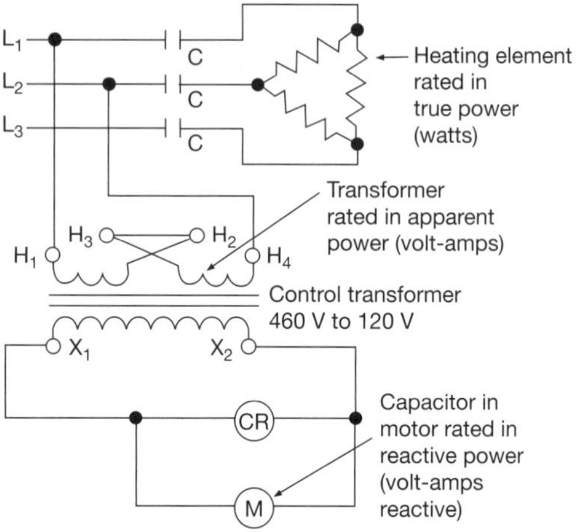

Reactive power is supplied to a reactive load (capacitor/coil) and is measured in volt-amps reactive (VAR). The capacitor on a motor uses reactive power to keep the capacitor charged. The capacitor uses no true power because it performs no actual work such as producing heat or motion.

In an AC circuit containing only resistance, the power in the circuit is true power. However, almost all AC circuits include capacitive reactance (capacitors) and/or inductive reactance (coils). Inductive reactance is the most common, because all motors, transformers, solenoids, and coils have inductive reactance.

Apparent power represents a load or circuit that includes both true power and reactive power and is expressed in volt-amps (VA), kilovolt-amps (kVA) or megavolt-amps (MVA). Apparent power is a measure of component or system capacity because apparent power considers circuit current regardless of how it is used. For this reason, transformers are sized in volt-amps rather than in watts.

TRUE POWER AND APPARENT POWER

True power is the actual power used in an electrical circuit and is expressed in watts (W). *Apparent power* is the product of voltage and current in a circuit calculated without considering the phase shift that may be present between total voltage and current in the circuit. Apparent power is measured in volt-amperes (VA). A phase shift exists in most AC circuits that contain devices causing capacitance or inductance.

True power equals apparent power in an electrical circuit containing only resistance. True power is less than apparent power in a circuit containing inductance or capacitance.

Capacitance is the property of a device that permits the storage of electrically separated charges when potential differences exist between the conductors. *Inductance* is the property of a circuit that causes it to oppose a change in current due to energy stored in a magnetic field; i.e., coils.

To calculate true power, apply the formula:

$$P_T = I^2 \times R$$

where

P_T = true power (in watts)
I = total circuit current (in amps)
R = total resistive component of the circuit (in ohms)

To calculate apparent power, apply the formula:

$$P_A = E \times I$$

where

P_A = apparent power (in volt-amps)
E = measured voltage (in volts)
I = measured current (in amps)

POWER FACTOR FORMULA

Power factor is the ratio of true power used in an AC circuit to apparent power delivered to the circuit.

$$PF = \frac{P_T}{P_A}$$

where

PF = power factor (percentage)
P_T = true power (in watts)
P_A = apparent power (in volt-amps)

POWER FACTOR IMPROVEMENT

Capacitor Multipliers for Kilowatt Load

To give capacitor kvar required to improve power factor from original to desired value. For example, assume the total plant load is 100kw at 60 percent power factor. Capacitor kvar rating necessary to improve power factor to 80 percent is found by multiplying kw (100) by multiplier in table (0.583), which gives kvar (58.3). Nearest standard rating (60 kvar) should be recommended.

Original Power Factor Percentage	Desired Power Factor Percentage				
	100%	95%	90%	85%	80%
60	1.333	1.004	0.849	0.713	0.583
62	1.266	0.937	0.782	0.646	0.516
64	1.201	0.872	0.717	0.581	0.451
66	1.138	0.809	0.654	0.518	0.388
68	1.078	0.749	0.594	0.458	0.338
70	1.020	0.691	0.536	0.400	0.270
72	0.964	0.635	0.480	0.344	0.214
74	0.909	0.580	0.425	0.289	0.159
76	0.855	0.526	0.371	0.235	0.105
77	0.829	0.500	0.345	0.209	0.079
78	0.802	0.473	0.318	0.182	0.052
79	0.776	0.447	0.292	0.156	0.026
80	0.750	0.421	0.266	0.130	
81	0.724	0.395	0.240	0.104	
82	0.698	0.369	0.214	0.078	
83	0.672	0.343	0.188	0.052	
84	0.646	0.317	0.162	0.026	
85	0.620	0.291	0.136		
86	0.593	0.264	0.109		
87	0.567	0.238	0.083		
88	0.540	0.211	0.056		
89	0.512	0.183	0.028		
90	0.484	0.155			
91	0.456	0.127			
92	0.426	0.097			
93	0.395	0.066			
94	0.363	0.034			
95	0.329				
96	0.292				
97	0.251				
99	0.143				

CAPACITOR CORRECTION FOR THREE-PHASE MOTORS

	600 RPM		900 RPM		1200 RPM		1800 RPM		3600 RPM	
HP	Capacitor Rating KVAR	% of Current Reduced	Capacitor Rating KVAR	% of Current Reduced	Capacitor Rating KVAR	% of Current Reduced	Capacitor Rating KVAR	% of Current Reduced	Capacitor Rating KVAR	% of Current Reduced
10	7.5	31	5.0	21	3.5	14	3.0	11	3.0	10
15	9.5	27	6.5	18	5.0	13	4.0	10	4.0	9
20	12.0	25	7.5	16	6.5	12	5.0	10	5.0	9
25	14.0	23	9.0	15	7.5	11	6.0	10	6.0	9
30	16.0	22	10.0	14	9.0	11	7.0	9	7.0	8
40	20.0	20	12.0	13	11.0	10	9.0	9	9.0	8
50	24.0	19	15.0	12	13.0	10	11.0	9	12.0	8
60	27.0	19	18.0	11	15.0	10	14.0	8	14.0	8
75	32.5	18	21.0	10	18.0	10	16.0	8	17.0	8
100	40.0	17	27.0	10	25.0	9	21.0	8	22.0	8
125	47.5	16	32.5	10	30.0	9	26.0	8	27.0	8
150	52.5	15	37.5	10	35.0	9	30.0	8	32.5	8
200	65.0	14	47.5	10	42.5	9	37.5	8	40.0	8

Motors shown are open-type induction at 60 hertz.
For lower rated capacitors, reduce proportionately to line current.

DELTA AND WYE RESISTOR CIRCUITS

In the delta network the resistance between terminals may be determined by combining the formulas for series and parallel resistances as follows:

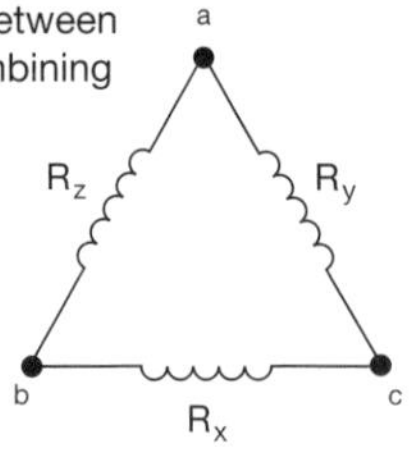

Delta

$$\text{a to b} = \frac{R_z \times (R_x + R_y)}{R_z + (R_x + R_y)}$$

$$\text{a to c} = \frac{R_y \times (R_x + R_z)}{R_y + (R_x + R_z)}$$

$$\text{b to c} = \frac{R_x \times (R_z + R_y)}{R_x + (R_z + R_y)}$$

To convert from delta to wye, use:

$$R_a = \frac{R_y \times R_z}{R_x + R_y + R_z}$$

$$R_b = \frac{R_x \times R_z}{R_x + R_y + R_z}$$

$$R_c = \frac{R_x \times R_y}{R_x + R_y + R_z}$$

Wye

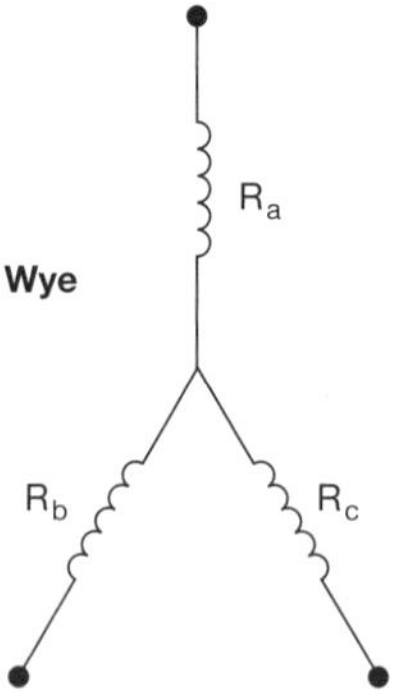

To convert from wye to delta, use:

$$R_x = \frac{(R_a \times R_b) + (R_b \times R_c) + (R_c \times R_a)}{R_a}$$

$$R_y = \frac{(R_a \times R_b) + (R_b \times R_c) + (R_c \times R_a)}{R_b}$$

$$R_z = \frac{(R_a \times R_b) + (R_b \times R_c) + (R_c \times R_a)}{R_c}$$

SUMMARY OF SERIES, PARALLEL, AND COMBINATION CIRCUITS			
To Find	**Series Circuits**	**Parallel Circuits**	**Series/Parallel**
Resistance (R)	$R_T = R_1 + R_2 + R_3$ Sum of individual resistances	$\frac{1}{R_T} = \frac{1}{R_1} + \frac{1}{R_2} + \frac{1}{R_3}$	Total resistance equals resistance of parallel portion and sum of series resistors
Current (I)	$I_T = I_1 = I_2 = I_3$ The same throughout entire circuit	$I_T = I_1 + I_2 + I_3$ Sum of individual currents	Series rules apply to series portion of circuit Parallel rules apply to parallel part of circuit
Voltage (E)	$E_T = E_1 + E_2 + E_3$ Sum of individual voltages	$E_T = E_1 = E_2 = E_3$ Total voltage and branch voltage are the same	Total voltage is sum of voltage drops across each series resistor and each of the branches of parallel portion
Power (P)	$P_T = P_1 + P_2 + P_3$ Sum of individual wattages	$P_T = P_1 + P_2 + P_3$ Sum of individual wattages	$P_T = P_1 + P_2 + P_3$ Sum of Individual wattages

CIRCUIT CHARACTERISTICS

Resistances in a Series DC Circuit

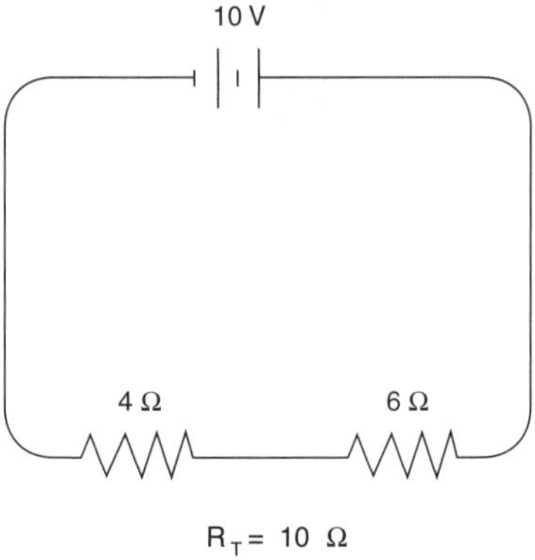

Parallel Circuit, Showing Voltage Drops

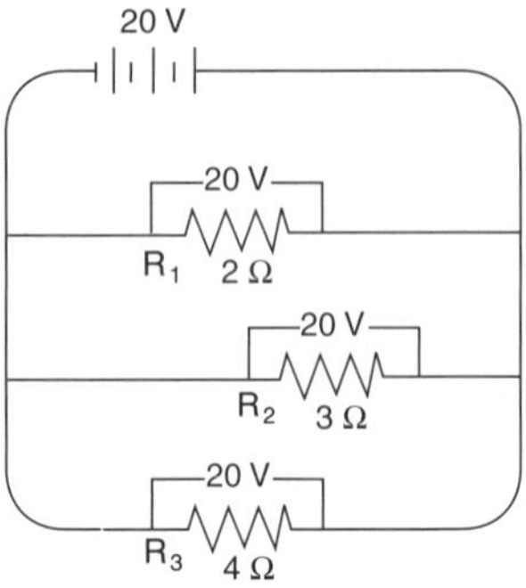

Voltages in a Series-Parallel Circuit

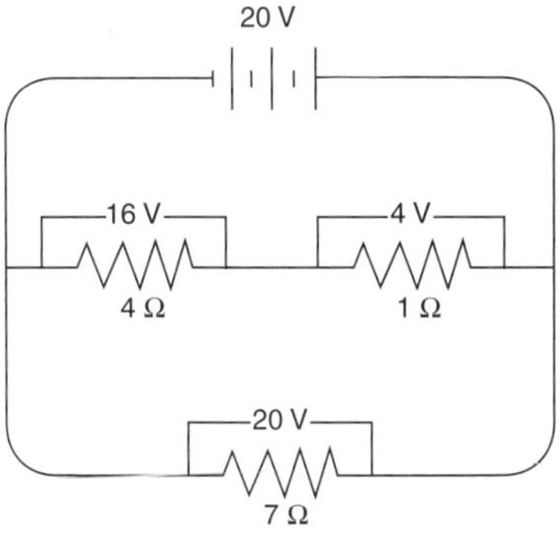

Parallel Circuit, Showing Current Values

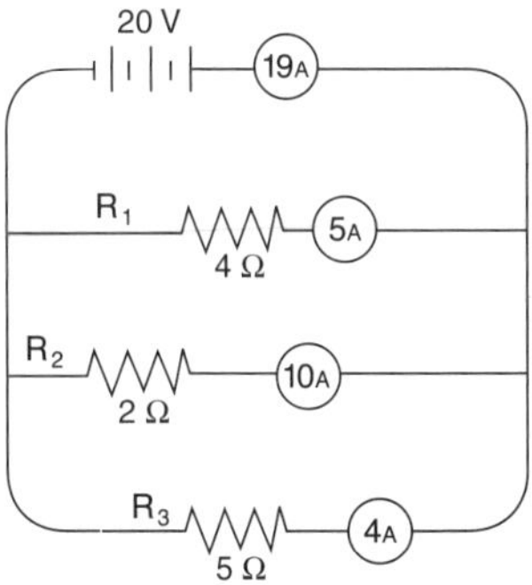

$I_1 = 5_A$

$I_2 = 10_A$

$I_3 = 4_A$

$I_T = 19_A$

CAPACITANCE

Capacitors in Series

$$C_T = \frac{C_1 \times C_2}{C_1 + C_2} \quad \text{(Two only)}$$

$$\frac{1}{C_T} = \frac{1}{C_1} + \frac{1}{C_2} + \frac{1}{C_3} + \cdots \quad \text{(Multiple)}$$

Capacitors in Parallel

$$C_T = C_1 + C_2 + C_3 + \cdots$$

CAPACITORS CONNECTED IN SERIES/PARALLEL

1. Calculate the capacitance of the parallel branch.

$$C_{PT} = C_1 + C_2 + C_3 + \cdots$$

2. Calculate the capacitance of the series combination.

$$C_T = \frac{C_{PT} \times C_S}{C_{PT} + C_S}$$

CAPACITIVE REACTANCE

Capacitive reactance (X_C) is the opposition to current flow by a capacitor when connected to an AC power supply and is expressed in ohms. To calculate capacitive reactance, apply the formula:

$$X_C = \frac{1}{2\pi fc}$$

where

X_C = capacitive reactance (in Ohms)

2π = 6.28 (indicates circular motion that produces the AC sine wave)

f = applied frequency (in Hertz)

c = capacitance (in Farads)

Capacitive Reactances

In Series

$$X_{CT} = X_{C1} + X_{C2} + X_{C3} \cdots$$

In Parallel

$$X_{CT} = \frac{X_{C_1} \times X_{C_2}}{X_{C_1} + X_{C_2}} \quad \text{(Two Only)}$$

(Multiple)

$$\frac{1}{X_{C_T}} = \frac{1}{X_{C_1}} + \frac{1}{X_{C_2}} + \frac{1}{X_{C_3}} + \cdots$$

To calculate capacitive reactance when voltage across the capacitor (E_C) and current through the capacitor (I_C) are known, apply the formula:

$$X_C = \frac{E_C}{I_C}$$

where

X_C = capacative reactance (in Ohms)

E_C = voltage across capacitor (in Volts)

I_C = current through capacitor (in Amps)

INDUCTANCE

Inductors in Series

$$L_T = L_1 + L_2 + L_3 + \cdots$$

Inductors in Parallel

$$L_T = \frac{L_1 \times L_2}{L_1 + L_2} \quad \text{(Two Only)}$$

$$\frac{1}{L_T} = \frac{1}{L_1} + \frac{1}{L_2} + \frac{1}{L_3} + \cdots \quad \text{(Multiple)}$$

INDUCTORS CONNECTED IN SERIES/PARALLEL

(Two Only)

parallel — series

$$L_T = \left(\frac{L_1 \times L_2}{L_1 + L_2}\right) + L_1 + L_2 + L_3 + \cdots$$

(Multiple)

parallel — series

$$L_T = \left(\frac{1}{\frac{1}{L_1} + \frac{1}{L_2} + \frac{1}{L_3} + \cdots}\right) + L_1 + L_2 + L_3 + \cdots$$

INDUCTIVE REACTANCE

Inductive reactance (X_L) is an inductor's opposition to alternating current measured in ohms.

To calculate inductive reactance for AC circuits, apply the formula:

$$X_L = 2\pi fL$$

where

X_L = inductive reactance (in Ohms)
2π = 6.28 (indicates circular motion that produces the AC sine wave)
f = applied frequency (in Hertz)
L = inductance (in Henrys)

Inductive Reactances

In Series

$$X_{L_T} = X_{L_1} + X_{L_2} + \cdots$$

In Parallel

$$X_{LT} = \frac{X_{L_1} \times X_{L_2}}{X_{L_1} + X_{L_2}} \quad \text{(Two Only)}$$

$$\frac{1}{X_{L_T}} = \frac{1}{X_{L_1}} + \frac{1}{X_{L_2}} + \frac{1}{X_{L_3}} \cdots \quad \text{(Multiple)}$$

To calculate inductive reactance when voltage across a coil (E_L) and current through a coil (I_L) are known.

$$X_L = \frac{E_L}{I_L}$$

where

X_L = inductive reactance (in Ohms)
E_L = voltage across coil (in Volts)
I_L = current through coil (in Amps)

VOLTAGE DROP USING OHM'S LAW

The definition of OHM's LAW is: Voltage equals current (Amperes) times resistance (Ohms).

A working model using the following criteria will calculate the voltage drop in a power line. The line is 300 feet long (150' and back) using 12 gauge copper wire powering an 800 Watt light fixture from a 120 Volt supply.

Resistance of #12 wire is 0.00162 Ohms/Foot

Current is 800 Watts/120 Volts = 6.67 Amps

Resistance is 300 Feet × .00162 Ohms/Foot = 0.486 Ohms

Voltage Drop = 6.67 Amps × 0.486 Ohms = 3.24 Volts

The Voltage Drop in percentage = 3.24 Volts/120 Volts = 2.7%

USING DISTANCE TO LOAD

For a 1ϕ, 2-wire, line-to-line or line-to-neutral	Use $E = I \times R \times 2$
For a 1ϕ, 3-wire, line-to-neutral with a balanced neutral	Use $E = I \times R$
For a 3ϕ, 3-wire, line-to-line	Use $E = I \times R \times 1.73$
For a 3ϕ, 4-wire, line-to-neutral with a balanced neutral	Use $E = I \times R$

E = drop in volts (to calculate in %, divide E by the circuit voltage).
I = amperage of the load.
R = resistance of the conductors based on length in one direction.

CALCULATING BRANCH CIRCUIT VOLTAGE DROP IN PERCENT

$$\% V_D = \frac{V_{NL} - V_{FL}}{V_{NL}} \times 100$$

where $\% V_D$ = percent voltage drop
V_{NL} = no-load voltage (Volts)
V_{FL} = full-load voltage (Volts)

VOLTAGE DROP FORMULAS

The NEC® recommends a maximum 3% voltage drop for either the branch circuit or the feeder.

Single-phase:

$$VD = \frac{2 \times R \times I \times L}{CM}$$

Three-phase:

$$VD = \frac{1.732 \times R \times I \times L}{CM}$$

VD = Volts (voltage drop of the circuit)

R = 12.9 ohms/copper or 21.2 ohms/aluminum (resistance constants for a 1,000 circular mils conductor that is 1,000 ft. long, at an operating temperature of 75°C).

I = Amps (load at 100 %)

L = Ft. (length of circuit from load to power supply)

CM = Circular-mils (conductor wire size)

2 = Single-phase constant

1.732 = Three-phase constant

CONDUCTOR LENGTH/VOLTAGE DROP

Voltage drop can be reduced by limiting the length of the conductors.

Single-phase:

$$L = \frac{CM \times VD}{2 \times R \times I}$$

Three-phase:

$$L = \frac{CM \times VD}{1.732 \times R \times I}$$

CONDUCTOR SIZE/VOLTAGE DROP

Increase the size of the conductor to decrease the voltage drop of circuit (reduce its resistance).

Single-phase:

$$CM = \frac{2 \times R \times I \times L}{VD}$$

Three-phase:

$$CM = \frac{1.732 \times R \times I \times L}{VD}$$

WIRE LENGTH VS WIRE SIZE (MAX. VOLTAGE DROP)

Max. Wire Ft. @ 240 V, Single-Phase, 2% Max. Voltage Drop

3/0	2/0	1/0	**#2**	**#4**	**Watts**	**Amps**
180	—	—	—	—	48,000	200
240	190	185	—	—	36,000	150
360	280	230	—	—	24,000	100
440	365	290	180	—	19,200	80
520	415	330	205	130	16,800	70
600	485	385	240	150	14,400	60
720	580	460	290	180	12,000	50
880	725	575	360	230	5,600	40
1,200	970	770	485	300	7,200	30
1,440	1,100	920	580	365	6,000	25

#6	**#8**	**#10**	**#12**	**#14**	**Watts**	**Amps**
105	—	—	—	—	12,000	50
130	90	—	—	—	5,600	40
175	120	75	—	—	7,200	30
210	144	90	—	—	6,000	25
265	180	110	70	—	4,800	20
350	240	150	95	60	3,600	15
525	360	225	140	90	2,400	10
1,020	720	455	285	180	1,200	5

Max. Wire Ft. @ 120 V, Single-Phase, 2% Max. Voltage Drop

3/0	2/0	1/0	**#2**	**#4**	**Watts**	**Amps**
230	180	144	90	—	9,600	80
260	205	165	105	65	8,400	70
305	240	190	120	76	7,200	60
360	290	230	145	90	6,000	50
440	360	290	175	115	4,800	40
600	490	385	240	150	3,600	30
720	580	460	290	180	3,000	25
900	725	575	365	230	2,400	20
1,200	965	770	485	305	1,800	15

#6	**#8**	**#10**	**#12**	**#14**	**Watts**	**Amps**
57	—	—	—	—	6,000	50
72	45	—	—	—	4,800	40
95	60	38	—	—	3,600	30
115	72	45	—	—	3,000	25
140	90	57	36	—	2,400	20
190	120	75	47	30	1,800	15
285	180	115	70	45	1,200	10
575	360	225	140	90	600	5

VOLTAGE DROP AMPERE-FEET

Copper Conductors, 70°C Copper Temp.,
600 V Class Single Conductor Cables in Steel Conduit

Conductor Size	Maximum circuit ampere-ft. without exceeding specified percentage voltage drop, various circuit voltages, and power factors.					
	DC Circuits 1% Drop on 120 V	60 Cycle AC Circuits				
		1% Drop, 1.00 PF		3% Drop, 0.85 PF		
		120 V 1-Phase	208 V 3-Phase	115 V 1-Phase	208 V 3-Phase	220 V 3-Phase
14	191	191	382	623	1,300	1,380
12	305	305	612	998	2,080	2,200
10	484	484	968	1,580	3,280	3,470
8	770	770	1,540	2,450	5,110	5,410
6	1,200	1,190	2,380	3,800	7,850	8,310
4	1,900	1,890	3,780	5,750	12,000	12,700
2	3,030	2,970	5,950	8,620	18,000	19,100
1	3,820	3,710	7,430	10,400	21,800	23,100
1/0	4,820	4,560	9,120	12,500	26,000	27,500
2/0	6,060	5,610	11,200	14,800	30,800	32,700
3/0	7,650	6,940	13,900	16,900	35,200	37,100
4/0	9,760	8,520	17,100	20,200	42,100	44,600
250	11,500	9,930	19,800	22,300	46,500	49,300
300	13,800	11,500	23,100	24,800	52,000	55,000
350	16,100	13,200	26,300	27,000	56,200	59,500
400	18,400	14,000	28,100	28,700	60,000	63,500
500	22,700	16,400	32,800	31,800	66,400	70,300
750	34,600	20,400	40,800	36,700	76,600	81,200

Note: Length to be used is the distance from point of supply to load.

VOLTAGE DROP TABLE

Conductor Size	Voltage Drop per 1000 Ampere-Feet							
		AC System						
		Load Power Factor (%)						
	DC	100	95	90	85	80	75	70
For DC circuit or single phase, 60-cycle, 2-wire system or 3-wire system with balanced load. Copper conductors, 70°C copper temperature. 600 V class single-conductor cables in steel conduit.								
14	6.29	6.29	6.06	5.78	5.54	5.26	4.97	4.74
12	3.93	3.93	3.81	3.64	3.46	3.29	3.13	2.95
10	2.48	2.48	2.44	2.31	2.19	2.08	1.96	1.85
8	1.56	1.56	1.51	1.47	1.41	1.34	1.27	1.20
6	0.999	1.011	0.987	0.953	0.918	0.872	0.826	0.774
4	0.631	0.635	0.641	0.624	0.600	0.578	0.554	0.528
2	0.396	0.404	0.418	0.413	0.400	0.386	0.372	0.358
1	0.314	0.323	0.356	0.337	0.330	0.322	0.311	0.300
1/0	0.249	0.263	0.280	0.282	0.277	0.269	0.263	0.255
2/0	0.198	0.214	0.233	0.236	0.233	0.230	0.226	0.222
3/0	0.157	0.173	0.196	0.206	0.204	0.200	0.194	0.188
4/0	0.123	0.141	0.163	0.170	0.171	0.170	0.169	0.166
250	0.1041	0.121	0.146	0.152	0.155	0.155	0.155	0.154
300	0.0870	0.1040	0.128	0.135	0.139	0.140	0.141	0.141
350	0.0746	0.0912	0.117	0.125	0.128	0.131	0.131	0.131
400	0.0652	0.0855	0.1086	0.117	0.120	0.122	0.124	0.125
500	0.0528	0.0733	0.0959	0.1040	0.1086	0.111	0.113	0.114
750	0.0347	0.0589	0.0808	0.0884	0.0940	0.0976	0.0999	0.1020
For 3-phase, 60-cycle, 3-wire or 4-wire balanced system. Copper conductors, 70°C copper temperature, 600 V class single-conductor cables in steel conduit.								
14		5.45	5.25	5.00	4.80	4.55	4.30	4.10
12		3.40	3.30	3.15	3.00	2.85	2.70	2.55
10		2.15	2.10	2.00	1.90	1.80	1.70	1.60
8		1.35	1.31	1.27	1.22	1.16	1.10	1.04
6		0.875	0.855	0.825	0.795	0.755	0.715	0.670
4		0.550	0.555	0.540	0.520	0.500	0.480	0.457
2		0.350	0.362	0.358	0.346	0.334	0.322	0.310
1		0.280	0.308	0.292	0.286	0.279	0.269	0.260
1/0		0.228	0.242	0.244	0.240	0.233	0.228	0.221
2/0		0.185	0.202	0.204	0.202	0.199	0.196	0.192
3/0		0.150	0.170	0.178	0.177	0.173	0.168	0.163
4/0		0.122	0.141	0.147	0.148	0.147	0.146	0.144
250		0.105	0.126	0.132	0.134	0.134	0.134	0.133
300		0.0900	0.111	0.117	0.120	0.121	0.122	0.122
350		0.0790	0.101	0.108	0.111	0.113	0.114	0.114
400		0.0740	0.0940	0.101	0.104	0.106	0.107	0.108
500		0.0635	0.0830	0.0900	0.0940	0.0964	0.0974	0.0988
750		0.0510	0.0700	0.0765	0.0814	0.0845	0.0865	0.0883

Note: Length to be used is the distance from point of supply to load, not amount of wire in circuit.

CHAPTER 2
Conduit, Cable and Underground Installations

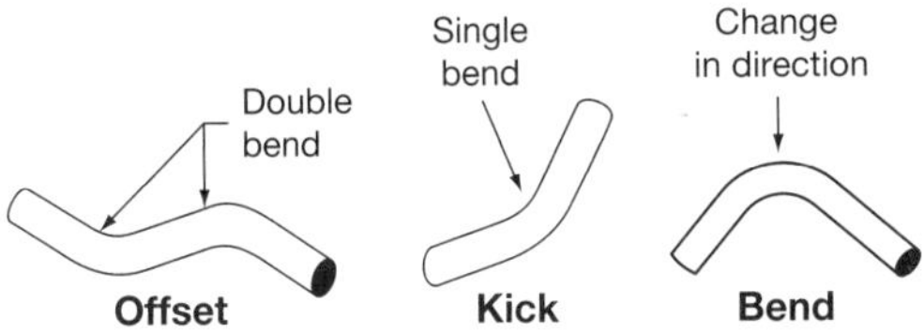

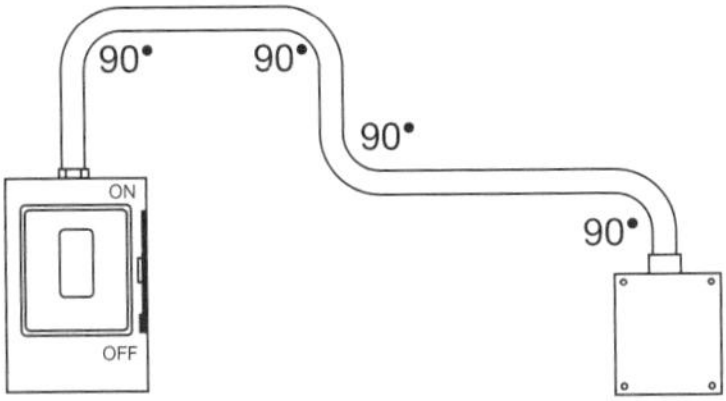

Total of 360° of bends is permitted.

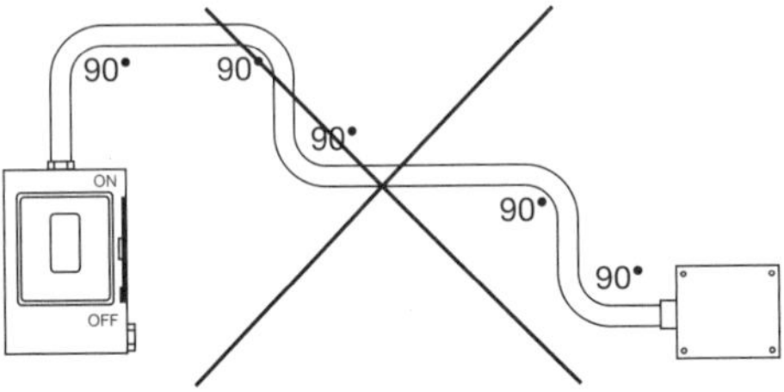

Over 360° of bends is not permitted.

BENDING EMT STUB-UPS	
EMT Size	**Take-up Amount**
½"	5"
¾"	6"
1"	8"
1¼"	11"

Step One:
Determine the height of the stub-up required, and mark on the EMT.

Step Two:
Subtract the take-up amount from the stub height, and mark the EMT that distance from the end.

Step Three:
Align the arrow on bender with the last mark made on the EMT, and bend to the 90° mark on the bender.

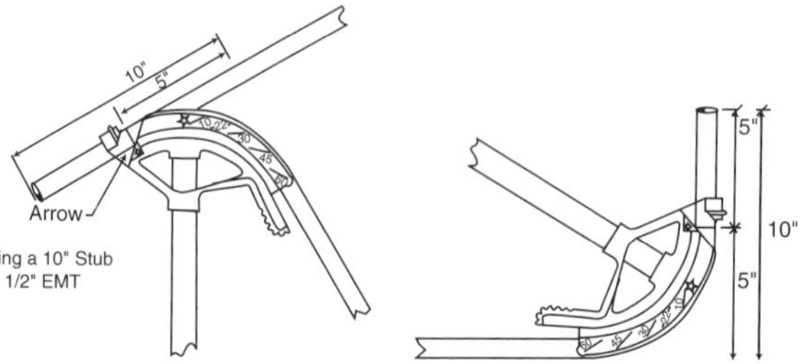

Making a 10" Stub from 1/2" EMT

BACK-TO-BACK BENDING

A back-to-back bend results in a "U" shaped length of conduit, usually for runs along the floor or ceiling which then turn up or down a wall.

Step One:
After the first 90° bend is made, determine the back-to-back length and mark it on the EMT.

Step Two:
Align this back-to-back mark with the star on the bender, and bend to 90°.

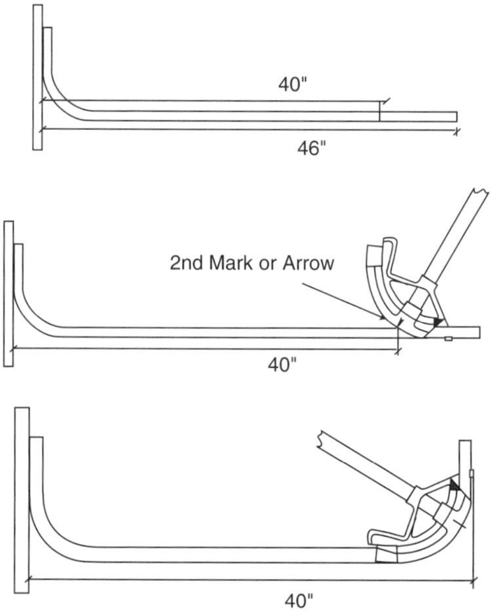

2-BEND OFFSET

2-Bend Offset

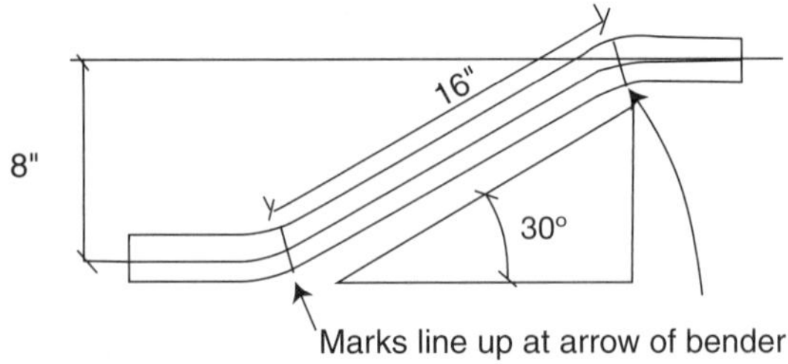

Angle of Bends	Inches of Run Per Inch of Offset	Loss of Conduit Length Per Inch of Offset
10°	5.76	1/16"
22.5°	2.6	3/16"
30°	2.0	1/4"
45°	1.414	3/8"
60°	1.15	1/2"

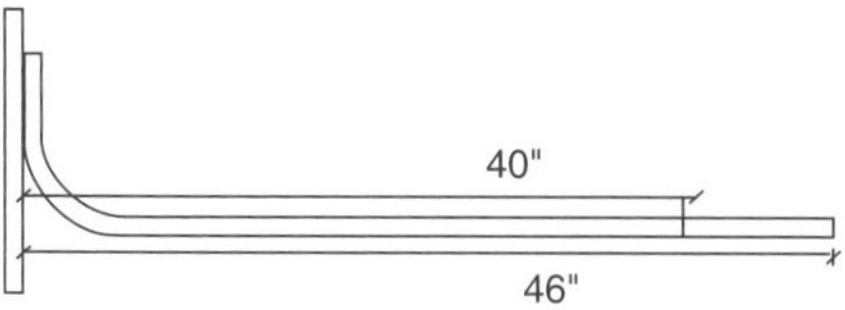

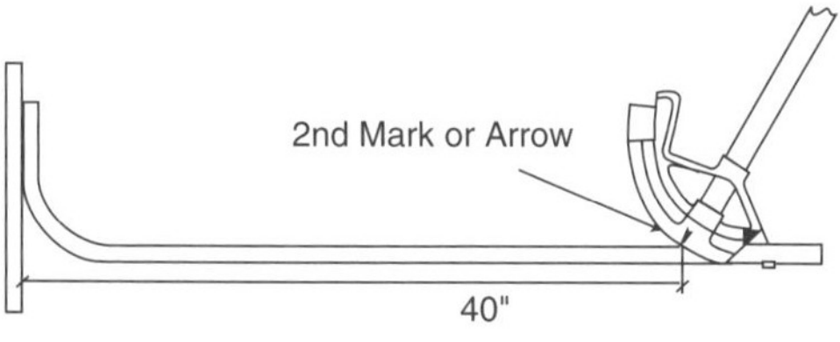

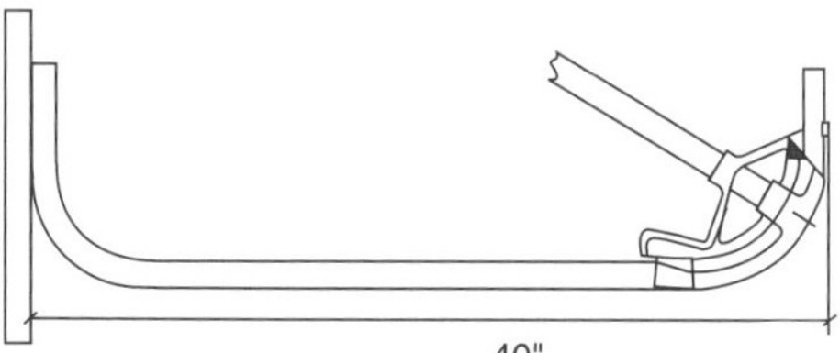

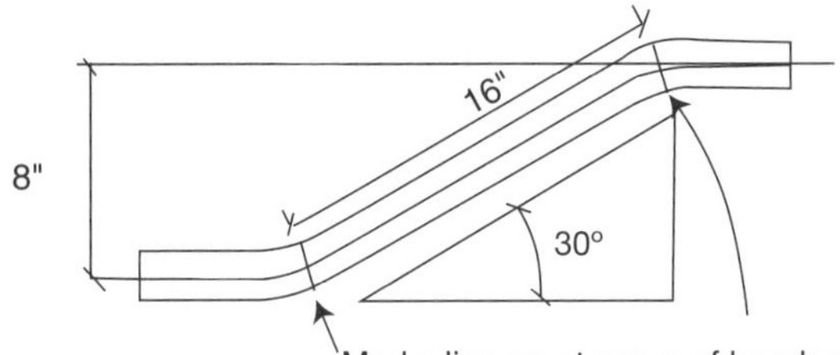

An offset bend is used to change the level, or plane, of the conduit, usually because of an obstruction in the original conduit path.

Step One:
Determine the offset depth.

Step Two:
Multiply the offset depth by the multiplier for the degree of bend used to determine the distance between bends. 30° is the usual choice, because of the convenient 2.0 multiplier.

For example, if the offset depth required is 8", and you intend to use 30° bends, the distance between bends is 8" × 2 = 16".

Step Three:
Mark at the appropriate points, align the arrow on the bender with the first mark, and bend to desired degree by aligning the EMT with chosen degree line on the bender.

Step Four:
Slide down the EMT, align the arrow with the second mark, and bend to the same degree line. Be aware of the position of the bender head.

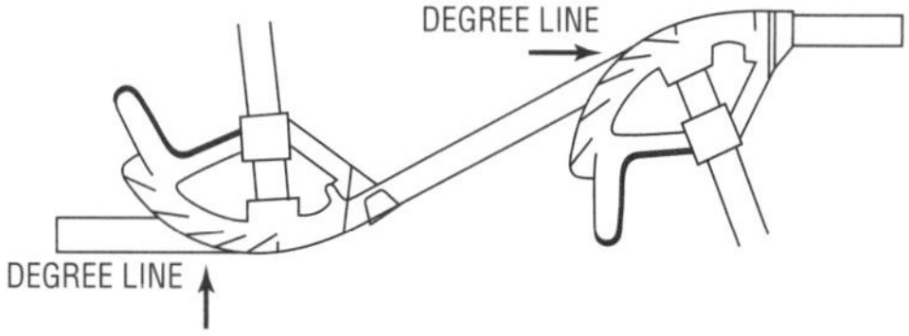

3-BEND SADDLE

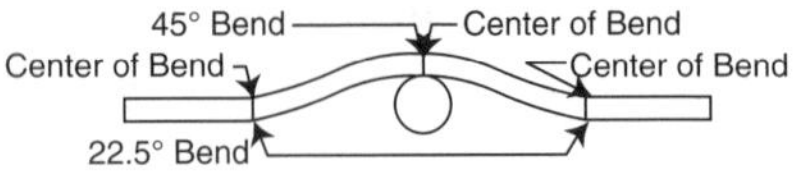

The 3-bend saddle is used when encountering an obstacle.

Step One:
Measure the height of the obstruction.
Mark the center point on the EMT.

Step Two:
Multiply the height of the obstruction by 2.5 and mark this distance on each side of the center mark, shown as D_2 and D_3.

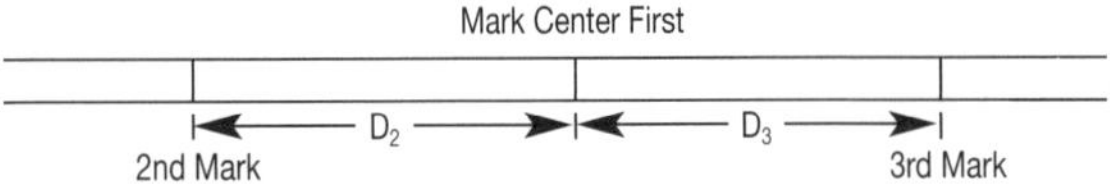

Step Three:
Place the center mark on the saddle mark/notch. Bend to 45°.

Step Four:
Bend the second mark to 22-½° angle at arrow.

Step Five:
Bend the third mark to 22-½° angle at arrow. Be sure to note the position of the EMT on all bends.

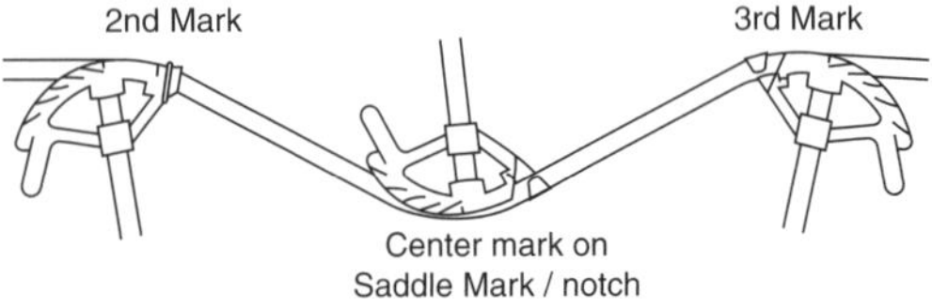

CHICAGO-TYPE CONDUIT BENDER — 90° BENDING

A to C = STUB-UP
C to D = TAIL
C = BACK OF STUB-UP
C = BOTTOM OF CONDUIT

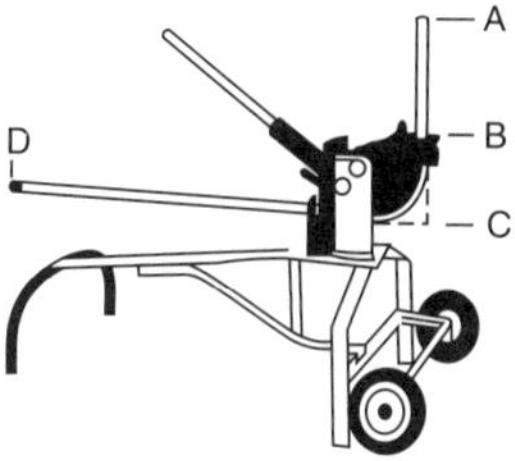

There are normally two sizes of bending shoes. The *small* shoe takes ½", ¾" and 1" conduit. The *large* shoe takes 1¼" and 1½" conduit.

To determine the "take-up" and "shrink" of each size conduit for any bender to make 90° bends, use a straight piece of conduit and:

1) Measure length of conduit, A to D.
2) Place conduit in bender and mark at the edge of the shoe, B.
3) Level conduit. Bend ninety, and count number of pumps. Be sure to note each size of conduit you have used.
4) After bending you can find:
   A. Distance between B and C is the take-up.
   B. Original measurement of the piece of conduit subtracted from distance A to C plus distance C to D is the shrink.

Notes: Time can be saved if conduit can be cut, reamed and threaded before you begin bending. Also, this same method can be used on hydraulic benders as well.

CHICAGO-TYPE CONDUIT BENDER — OFFSET BENDING

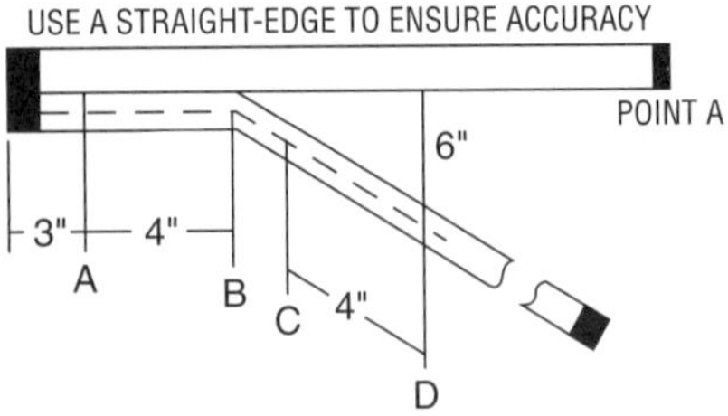

In order to bend a 6" off-set:

1. Make a mark 3" from the conduit end. Place the conduit in the bender with the mark at the outside edge of the jaw.
2. Make three full pumps, making sure handle goes all the way down.
3. Remove the conduit from the bender and place it alongside a straight-edge.
4. Measure 6" from the straight-edge to the center of the conduit. Mark it point D. Use a square to ensure accuracy.
5. Mark the center of the conduit from both directions through bend as shown by the broken line. Where the lines intersect is point B.
6. Measure from A to B to determine the distance from D to C. Mark C and place the conduit in the bender with the mark at the outside edge of the jaw, and with the kick pointing down. Use a level to ensure accuracy.
7. Make three full pumps, making sure the handle goes all the way down.

MULTI-SHOT 90° CONDUIT BENDING

Always measure, thread, cut and ream conduit before bending. The information contained in the following pages will take you through an example of multi-shot 90° conduit bending in a step by step process. The parameters of the example are as follows:

Size of the conduit = 2"
Space between conduits (center to center) = 6"
Height of the stub-up = 36"
Length of the tail = 48"

1) Determine the Radius (R):
   Conduit #1 will use the minimum radius which is 8 times the size of the conduit.
   Thus: Radius of Conduit #1 = 8 × 2" + 1.25" = 17.25"
   Radius of Conduit #2 = R_1 + 6" = 23.25"
   Radius of Conduit #3 = R_2 + 6" = 29.25"
2) Determine the Developed Length (DL):
   Radius × 1.57 = DL
   Thus: DL of Conduit #1 = R_1 × 1.57 = 17.25" × 1.57 = 27"
   DL of Conduit #2 = R_2 × 1.57 = 23.25" × 1.57 = 36.5"
   DL of Conduit #3 = R_3 × 1.57 = 29.25" × 1.57 = 46"
3) Determine the Length of the Nipple:
   Thus: Nipple Length of Conduit #1
   = (L_1 + H_1 + DL_1) − (2 × R_1)
   = (48" + 36" + 27") − (34.5") = 76.5"
   Nipple Length of Conduit #2 = (L_2 + H_2 + DL_2) − (2 × R_2)
   = (54" + 42" + 36.5") − (46.5") = 86"
   Nipple Length of Conduit #3 = (L_3 + H_3 + DL_3) − (2 × R_3)
   = (60" + 48" + 46") − (58.5") = 95.5"

Note: For 90° bends, the shrink = (2 × R) − DL
For offset bends, the shrink = Hypotenuse − Side Adjacent

MULTI-SHOT 90° CONDUIT BENDING *(cont.)*

For the Layout and the bending, follow the procedures below:

1) Locate point B by measuring from point A, the Height of the stub-up (H_1) minus the radius (R_1). This distance would be 18.75" on all three conduits.
2) Locate point C by measuring from point D, the Length (L_1) minus the radius (R_1). This distance would be 30.75" on all three conduits.
3) Divide the Developed Length (DL), which is point B to point C, into equal spaces which should not be greater than 1.75" in order to prevent any possible wrinkling of the conduit. On Conduit #1, 17 spaces of 1.5882" each would result in 18 shots of 5° each. Note that there is always one less space than there are shots. When you are determining the quantity of shots, select a number that will divide evenly into 90.
4) An alternative method using a table corner and a ruler can be utilized as shown below:

A to B = Conduit #1
Developed Length or
DL (B to C) = 27"

A to C = 17 1" spaces
A to B = 17 1.5882" spaces
C = table corner

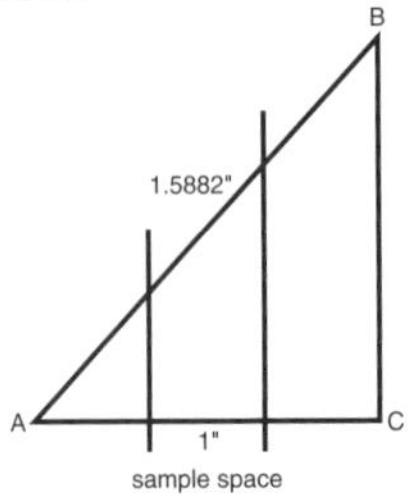

Measure from Point C (table corner) 17 inches along table edge to Point A and mark. Place end of rule at Point A. Point B will be located where 27" mark meets table edge B–C. Mark on board, then transfer to conduit.

MULTI-SHOT 90° CONDUIT BENDING *(cont.)*

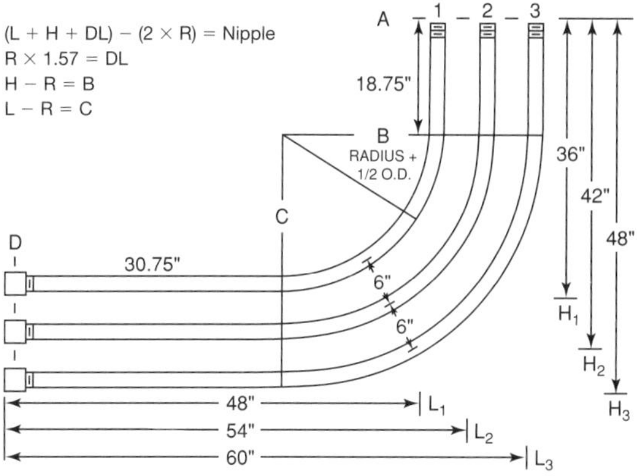

TO LOCATE POINT B

$B = H_1 - R_1$
$= 36" - 17.25"$
$= 18.75"$

$B = H_2 - R_2$
$= 42" - 23.25"$
$= 18.75"$

$B = H_3 - R_3$
$= 48" - 29.25"$
$= 18.75"$

TO LOCATE POINT C

$C = L_1 - R_1$
$= 48" - 17.25"$
$= 30.75"$

$C = L_2 - R_2$
$= 54" - 23.25"$
$= 30.75"$

$C = L_3 - R_3$
$= 60" - 29.25"$
$= 30.75"$

Note that Points B and C are equidistant from the end on all three conduits.

BOX OFFSETS					
Bend Angle	**Center-to-Center Distance**				
	1/4"	3/8"	1/2"	5/8"	3/4"
5°	2 7/8"	4 5/16"	5 3/4"	7 3/16"	8 5/8"
10°	1 7/16"	2 3/16"	2 7/8"	3 9/16"	4 5/16"

OFFSET TRAVEL					
Bend Angle	**Travel in Inches**				
	1/2"	3/4"	1"	1 1/4"	1 1/2"
5°	1/4" +	7/16" −	9/16"	3/4"	1"
10°	1/2" −	1 1/16"	1 3/16"	1 3/8"	1 11/16"
15°	11/16"	1 1/4" +	1 9/16"	2 3/16"	2 5/8"
20°	15/16"	1 5/8"	2 1/8"	2 13/16"	3 7/16"
22.5°	1 1/16"	1 7/8"	2 3/8"	3 3/16"	3 11/16"
30°	1 7/16"	2 7/16" +	3 1/16"	4 1/8"	4 3/4"
40°	1 7/8"	3 5/16"	4 1/8"	5 3/8"	5 1/2"
45°	2 1/8"	3 5/8"	4 3/8"	6"	6 1/4"
60°	2 13/16" +	4 7/8"	6 1/4"	8 1/8"	9 3/8"
90°	4 3/16"	7 7/16"	9 3/8"	12 1/16"	13 13/16"

KICK ADJUSTMENTS					
Bend Angle	**Adjustment in Inches**				
	1/2"	3/4"	1"	1 1/4"	1 1/2"
20°	1 19/32"	2 13/32"	2 11/16"	2 13/16"	3 3/8"
22.5°	1 5/8"	2 1/2"	2 19/32"	3"	3 5/8"
30°	1 27/32"	2 11/32"	3 17/32"	3 9/16"	4 5/32"
45°	2 7/32"	3 19/32"	4 5/16"	4 5/8"	5 17/32"
60°	2 23/32"	4 3/8"	5 13/32"	5 3/4"	7"

EMT TAKE-UP FOR HAND BENDERS

Hand Bender	Take-up
½"	5"
¾"	6"
1"	8"
1¼"	12"

GAIN AND RADIUS

Conduit Size	Gain	Radius
½"	2⅝"	4"
¾"	3¼"	5"
1"	4"	6"
1¼"	5⅝"	8"

SUPPORT SPACING FOR RIGID METAL CONDUIT

Conduit Size	Max. Distance Between Supports
½" to ¾"	10'
1"	12'
1¼" to 1½"	14'
2" to 2½"	16'
3" and larger	20'

SUPPORT SPACING FOR EMT AND METAL CONDUIT

All sizes: Every 10', plus within 36" of every box or other termination.

SUPPORT SPACING FOR RIGID PVC	
Conduit Size	Max. Distance Between Supports
½" to 1"	3'
1¼" to 1"	5'
2½" to 3"	6'
3½" to 5"	7'
6"	8'

SUPPORT SPACING FOR FLEXIBLE CONDUITS
All sizes: Every 4'6", plus within 12" of every box or other termination.

EXPANSION CHARACTERISTICS OF PVC RIGID NONMETALLIC CONDUIT			
Temperature Change in Degrees F	Length Change in Inches per 100 Ft.	Temperature Change in Degrees F	Length Change in Inches per 100 Ft.
10°	0.41	110°	4.46
20°	0.81	120°	4.87
30°	1.22	130°	5.27
40°	1.62	140°	5.68
50°	2.03	150°	6.08
60°	2.43	160°	6.49
70°	2.84	170°	6.90
80°	3.24	180°	7.30
90°	3.65	190°	7.71
100°	4.06	200°	8.11

CONDUIT AND TUBING—ALLOWABLE AREA DIMENSIONS FOR WIRE COMBINATIONS

Trade Size Inches	Trade I.D. Inches	100% Total Area Sq. In.	31% 2 Wires Sq. In.	40% Over 2 Wires Sq. In.	53% 1 Wire Sq. In.
RIGID METAL CONDUIT					
½	0.632	0.314	0.097	0.125	0.166
¾	0.836	0.549	0.170	0.220	0.291
1	1.063	0.887	0.275	0.355	0.470
1¼	1.394	1.526	0.473	0.610	0.809
1½	1.624	2.071	0.642	0.829	1.098
2	2.083	3.408	1.056	1.363	1.806
2½	2.489	4.866	1.508	1.946	2.579
3	3.090	7.499	2.325	3.000	3.974
3½	3.570	10.010	3.103	4.004	5.305
4	4.050	12.882	3.994	5.153	6.828
5	5.073	20.212	6.266	8.085	10.713
6	6.093	29.158	9.039	11.663	15.454
LIQUIDTIGHT FLEXIBLE METAL CONDUIT					
⅜	0.494	0.192	0.059	0.077	0.102
½	0.632	0.314	0.097	0.125	0.166
¾	0.830	0.541	0.168	0.216	0.287
1	1.054	0.873	0.270	0.349	0.462
1¼	1.395	1.528	0.474	0.611	0.810
1½	1.588	1.981	0.614	0.792	1.050
2	2.033	3.246	1.006	1.298	1.720
2½	2.493	4.881	1.513	1.953	2.587
3	3.085	7.475	2.317	2.990	3.962
3½	3.520	9.731	3.017	3.893	5.158
4	4.020	12.692	3.935	5.077	6.727
LIQUIDTIGHT FLEXIBLE NONMETALLIC CONDUIT (TYPE LFNC-A)					
⅜	0.495	0.192	0.060	0.077	0.102
½	0.630	0.312	0.097	0.125	0.165
¾	0.825	0.535	0.166	0.214	0.283
1	1.043	0.854	0.265	0.342	0.453
1¼	1.383	1.502	0.466	0.601	0.796
1½	1.603	2.018	0.626	0.807	1.070
2	2.063	3.343	1.036	1.337	1.772
LIQUIDTIGHT FLEXIBLE NONMETALLIC CONDUIT (TYPE LFNC-B)					
⅜	0.494	0.192	0.059	0.077	0.102
½	0.632	0.314	0.097	0.125	0.166
¾	0.830	0.541	0.168	0.216	0.287
1	1.054	0.873	0.270	0.349	0.462
1¼	1.395	1.528	0.474	0.611	0.810
1½	1.588	1.981	0.614	0.792	1.050
2	2.033	3.246	1.006	1.298	1.720

CONDUIT AND TUBING—ALLOWABLE AREA DIMENSIONS FOR WIRE COMBINATIONS

Trade Size Inches	Trade I.D. Inches	100% Total Area Sq. In.	31% 2 Wires Sq. In.	40% Over 2 Wires Sq. In.	53% 1 Wire Sq. In.
ELECTRICAL NONMETALLIC TUBING					
½	0.560	0.246	0.076	0.099	0.131
¾	0.760	0.454	0.141	0.181	0.240
1	1.000	0.785	0.243	0.314	0.416
1¼	1.340	1.410	0.437	0.564	0.747
1¼	1.570	1.936	0.600	0.774	1.026
2	2.020	3.205	0.993	1.282	1.699
ELECTRICAL METALLIC TUBING					
½	0.622	0.304	0.094	0.122	0.161
¾	0.824	0.533	0.165	0.213	0.283
1	1.049	0.864	0.268	0.346	0.458
1¼	1.380	1.496	0.464	0.598	0.793
1½	1.610	2.036	0.631	0.814	1.079
2	2.067	3.356	1.040	1.342	1.778
2½	2.731	5.858	1.816	2.343	3.105
3	3.356	8.846	2.742	3.538	4.688
3½	3.834	11.545	3.579	4.618	6.119
4	4.334	14.753	4.573	5.901	7.819
INTERMEDIATE METAL CONDUIT					
½	0.660	0.342	0.106	0.137	0.181
¾	0.864	0.586	0.182	0.235	0.311
1	1.105	0.959	0.297	0.384	0.508
1¼	1.448	1.647	0.510	0.659	0.873
1½	1.683	2.225	0.690	0.890	1.179
2	2.150	3.630	1.125	1.452	1.924
2½	2.557	5.135	1.592	2.054	2.722
3	3.176	7.922	2.456	3.169	4.199
3½	3.671	10.584	3.281	4.234	5.610
4	4.166	13.631	4.226	5.452	7.224
FLEXIBLE METAL CONDUIT					
⅜	0.384	0.116	0.036	0.046	0.061
½	0.635	0.317	0.098	0.127	0.168
¾	0.824	0.533	0.165	0.213	0.283
1	1.020	0.817	0.253	0.327	0.433
1¼	1.275	1.277	0.396	0.511	0.677
1½	1.538	1.858	0.576	0.743	0.985
2	2.040	3.269	1.013	1.307	1.732
2½	2.500	4.909	1.522	1.963	2.602
3	3.000	7.069	2.191	2.827	3.746
3½	3.500	9.621	2.983	3.848	5.099
4	4.000	12.566	3.896	5.027	6.660

CONDUIT AND TUBING—ALLOWABLE AREA DIMENSIONS FOR WIRE COMBINATIONS

Trade Size Inches	Trade I.D. Inches	100% Total Area Sq. In.	31% 2 Wires Sq. In.	40% Over 2 Wires Sq. In.	53% 1 Wire Sq. In.
PVC CONDUIT - TYPE EB					
2	2.221	3.874	1.201	1.550	2.053
3	3.330	8.709	2.700	3.484	4.616
3½	3.804	11.365	3.523	4.546	6.023
4	4.289	14.448	4.479	5.779	7.657
5	5.316	22.195	6.881	8.878	11.763
6	6.336	31.530	9.774	12.612	16.711
RIGID PVC CONDUIT - TYPE A					
½	0.700	0.385	0.119	0.154	0.204
¾	0.910	0.650	0.202	0.260	0.345
1	1.175	1.084	0.336	0.434	0.575
1¼	1.500	1.767	0.548	0.707	0.937
1½	1.720	2.324	0.720	0.929	1.231
2	2.155	3.647	1.131	1.459	1.933
2½	2.635	5.453	1.690	2.181	2.890
3	3.230	8.194	2.540	3.278	4.343
3½	3.690	10.694	3.315	4.278	5.668
4	4.180	13.723	4.254	5.489	7.273
RIGID PVC SCHEDULE 40 CONDUIT (HDPE)					
½	0.602	0.285	0.088	0.114	0.151
¾	0.804	0.508	0.157	0.203	0.269
1	1.029	0.832	0.258	0.333	0.441
1¼	1.360	1.453	0.450	0.581	0.770
1½	1.590	1.986	0.616	0.794	1.052
2	2.047	3.291	1.020	1.316	1.744
2½	2.445	4.695	1.455	1.878	2.488
3	3.042	7.268	2.253	2.907	3.852
3½	3.521	9.737	3.018	3.895	5.161
4	3.998	12.554	3.892	5.022	6.654
5	5.016	19.761	6.126	7.904	10.473
6	6.031	28.567	8.856	11.427	15.141
RIGID PVC SCHEDULE 80 CONDUIT					
½	0.526	0.217	0.067	0.087	0.115
¾	0.722	0.409	0.127	0.164	0.217
1	0.936	0.688	0.213	0.275	0.365
1¼	1.255	1.237	0.383	0.495	0.656
1½	1.476	1.711	0.530	0.684	0.907
2	1.913	2.874	0.891	1.150	1.523
2½	2.290	4.119	1.277	1.647	2.183
3	2.864	6.442	1.997	2.577	3.414
3½	3.326	8.688	2.693	3.475	4.605
4	3.786	11.258	3.490	4.503	5.967
5	4.768	17.855	5.535	7.142	9.463
6	5.709	25.598	7.935	10.239	13.567

DIMENSIONS AND WEIGHTS OF RIGID STEEL CONDUIT

Nominal or Trade Size of Conduit (Inches)	Inside Diameter (Inches)	Outside Diameter (Inches)	Wall Thickness (Inches)	Length Without Coupling Ft. & Ins.	Minimum Weight of Ten Unit Lengths with Couplings Attached (Pounds)
¼	0.364	0.540	0.088	9–11½	38.5
⅜	0.493	0.675	0.091	9–11½	51.5
½	0.622	0.840	0.109	9–11¼	79.0
¾	0.824	1.050	0.113	9–11¼	105.0
1	1.049	1.315	0.133	9–11	153.0
1¼	1.380	1.660	0.140	9–11	201.0
1½	1.610	1.900	0.145	9–11	249.0
2	2.067	2.375	0.154	9–11	334.0
2½	2.469	2.875	0.203	9–10½	527.0
3	3.068	3.500	0.216	9–10½	690.0
3½	3.548	4.000	0.226	9–10½	831.0
4	4.026	4.500	0.237	9–10¼	982.0
5	5.047	5.563	0.258	9–10	1344.0
6	6.065	6.625	0.280	9–10	1770.0

Note: The tolerances are:
Length: ± ¼-inch (without coupling)
Outside Diameter + 1/64-inch or –1/32-inch for the 1½-inch and smaller sizes
± 1 percent for the 2-inch and larger sizes
Wall Thickness: – 12½ percent

DIMENSIONS OF THREADS FOR RIGID STEEL CONDUIT

Nominal or Trade Size of Conduit (Inches)	Threads per Inch	Pitch Diameter at End of Thread E_0 (Inches)	Length of Thread (Inches)	
		Taper ¾ inch per Foot	Effective L_2	Over-All L_4
¼	18	0.4774	0.40	0.59
⅜	18	0.6120	0.41	0.60
½	14	0.7584	0.53	0.78
¾	14	0.9677	0.55	0.79
1	11½	1.2136	0.68	0.98
1¼	11½	1.5571	0.71	1.01
1½	11½	1.7961	0.72	1.03
2	11½	2.2690	0.76	1.06
2½	8	2.7195	1.14	1.57
3	8	3.3406	1.20	1.63
3½	8	3.8375	1.25	1.68
4	8	4.3344	1.30	1.73
5	8	5.3907	1.41	1.84
6	8	6.4461	1.51	1.95

Note: The tolerances are:
Thread Length (L_4, Column 5): ± 1 thread
Pitch Diameter (Column 3): ± 1 turn is the maximum variation permitted from the gauging face of the working thread gauges. This is equivalent to ± 1-½ turns from basic dimensions, since a variation of ±½ turn from basic dimensions is permitted in working gauges.

DIMENSIONS AND WEIGHTS OF RIGID STEEL COUPLINGS

Nominal or Trade Size of Conduit (Inches)	Outside Diameter (Inches)	Minimum Length (Inches)	Minimum Weight (Pounds)
1/4	0.719	1 3/16	0.055
3/8	0.875	1 3/16	0.075
1/2	1.010	1 9/16	0.115
3/4	1.250	1 5/8	0.170
1	1.525	2	0.300
1 1/4	1.869	2 1/16	0.370
1 1/2	2.155	2 1/16	0.515
2	2.650	2 1/8	0.671
2 1/2	3.250	3 1/8	1.675
3	3.870	3 1/4	2.085
3 1/2	4.500	3 3/8	3.400
4	4.875	3 1/2	2.839
5	6.000	3 3/4	4.462
6	7.200	4	7.282

Note: The tolerances are:

Outside Diameter: –1 percent for the 1 1/4-inch and larger sizes.
–1/64-inch for sizes smaller than 1 1/4-inch.

No limit is placed on the plus tolerances given for this dimension.

DIMENSIONS AND WEIGHTS OF RIGID STEEL 90° ELBOWS AND NIPPLES

Nominal or Trade Size of Conduit (Inches)	Elbows		Nipples	
	Minimum Radius to Center of Conduit (Inches)	Minimum Straight Length L_a at Each End (Inches)	A	B
1/4	–	–	–	–
3/8	–	–	–	–
1/2	4	1 1/2	0.065	2
3/4	4 1/2	1 1/2	0.086	4
1	5 3/4	1 7/8	0.125	9
1 1/4	7 1/4	2	0.164	10
1 1/2	8 1/4	2	0.202	11
2	9 1/2	2	0.269	14
2 1/2	10 1/2	3	0.430	60
3	13	3 1/8	0.561	70
3 1/2	15	3 1/4	0.663	90
4	16	3 3/8	0.786	115
5	24	3 5/8	1.060	170
6	30	3 3/4	1.410	200

Each lot of 100 nipples shall weigh not less than the number of pounds determined by the formula:

$$W = 100\,LA - B$$

Where W = weight of 100 nipples in pounds L = length of one nipple in inches
A = weight of nipple per inch in pounds
B = weight in pounds, lost in threading 100 nipples

DIMENSIONS AND WEIGHTS OF EMT

Nominal or Trade Size of Tubing (Inches)	Outside Diameter (Inches)	Minimum Wall Thickness (Inches)	Length (Feet)	Minimum Weight per 100 Feet (Pounds)
⅜	0.577	0.040	10	23
½	0.706	0.040	10	28.5
¾	0.922	0.046	10	43.5
1	1.163	0.054	10	64
1¼	1.510	0.061	10	95
1½	1.740	0.061	10	110
2	2.197	0.061	10	140

Note: The tolerances are:

Length:	±	¼ inch
Outside Dia.:	±	.005-inch
Wall Thick.:	+	18 percent

DIMENSIONS OF EMT 90° ELBOWS

Nominal or Trade Size of Tubing (Inches)	Minimum Radius to Center of Tubing (Inches)	Minimum Straight Length Ls at Each End (Inches)
½	4	1½
¾	4½	1½
1	5¾	1⅞
1¼	7¼	2
1½	8¼	2
2	9½	2

CONDUIT MASTER BUNDLES IN TOTAL LENGTH AND WEIGHT

R.M.C.			E.M.T.		
Size (in.)	Lg. (ft.)	Wt. (lb.)	Size (in.)	Lg. (ft.)	Wt. (lb.)
½	2,500	2,050	½	7,000	2,100
¾	2,000	2,180	¾	5,000	2,300
1	1,250	2,013	1	3,000	2,010
1¼	900	1,962	1¼	2,000	2,020
1½	800	2,104	1½	1,500	1,740
2	600	2,100	2	1,200	1,776
2½	370	2,068	2½	610	1,318
3	300	2,181	3	510	1,341
3½	250	2,200	3½	370	1,291
4	200	2,060	4	300	1,179
5	150	2,100			
6	100	1,840			

I.M.C.			Sched. 40 PVC		
Size (in.)	Lg. (ft.)	Wt. (lb.)	Size (in.)	Lg. (ft.)	Wt. (lb.)
½	3,500	2,170	½	6,000	972
¾	2,500	2,100	¾	4,400	951
1	1,700	2,023	1	3,600	1,152
1¼	1,350	2,133	1¼	3,300	1,433
1½	1,100	2,134	1½	2,250	1,170
2	800	2,048	2	1,400	979
2½	370	1,632	2½	930	1,031
3	300	1,629	3	880	1,276
3½	240	1,510	3½	630	1,099
4	240	1,680	4	570	1,178
			5	380	1,065
			6	260	946

NONMETALLIC-SHEATHED CABLE (TYPES NM AND NMC)

- Covered by Article 334 of the NEC®.
- Referred to as Romex with sizes #14 to #2 AWG with copper conductors and an equipment grounding conductor of proper size.
- The grounding conductor may be bare or covered with green insulation, or green insulation with one or more yellow stripes.
- NM may be used for both exposed and concealed work in normally dry locations.
- NM and NMC may not be used as service-entrance cable, embedded in poured cement, concrete, or aggregate. See the NEC for structures more than three floors above grade.
- NMC may be used for both exposed and concealed work in dry, moist, damp or corrosive locations.
- NM and NMC cables must be secured within 12 inches of every cabinet, box or fitting and at least every 4½ feet elsewhere.
- NM and NMC cables can be damaged due to sharp bending, being pulled through bored holes, or by driving the staples too forcefully.

UNDERGROUND FEEDER AND BRANCH-CIRCUIT CABLE (TYPE UF)

- Covered by Article 340 of the NEC®.
- Sizes #14 to 4/0 AWG with copper conductors and an equipment grounding conductor of proper size.
- The grounding conductor may be bare or covered with green insulation, or green insulation with one or more yellow stripes.
- Type UF cable may be buried directly in the earth at a minimum depth of 24 inches. This may be reduced to 18 inches beneath a 2-inch concrete pad, or encased in a metal raceway, pipe, or other suitable protection.
- UF cable may not be used as service-entrance cable, embedded in poured cement, concrete or aggregate, or exposed to sunlight.

ARMORED CABLE (TYPE AC)

- Covered by Article 320 of the NEC®.
- Referred to as BX and contains a bonding strip.
- AC cable may also contain an equipment grounding conductor of proper size which may be bare or covered with green insulation, or green insulation with one or more yellow stripes.
- May be used for both exposed and concealed work in normally dry locations and may be embedded in plaster finish on brick or other masonry, except in damp or wet locations.
- May be run or fished in the air voids of masonry block or tile walls, except where such walls are exposed or subject to excessive moisture or dampness or are below grade line.
- May not be installed in damp or wet locations, or where exposed to damage.
- AC cables must be secured within 12 inches of every cabinet, box, or fitting and at least every 4½ feet elsewhere.
- AC cables can be damaged due to sharp bending.
- At each termination, a fiber bushing must be inserted between the armored sheath and the conductors. The grounding strip must be bent back over the fiber bushing and in intimate contact with the external armor and the cable clamp.

SERVICE-ENTRANCE CABLE (TYPES SE AND USE)

- Covered by Article 338 of the NEC®.
- Used for service entrances of 100 amps.
- May be used for electric ranges and may be called Range cable.
- Cable will also contain an equipment grounding conductor of proper size which may be bare or covered with green insulation, or green insulation with one or more yellow stripes.
- SE cables must be secured within 12 inches of every cabinet, box, or fitting and at least every 4½ feet elsewhere.

UNDERGROUND INSTALLATION REQUIREMENTS

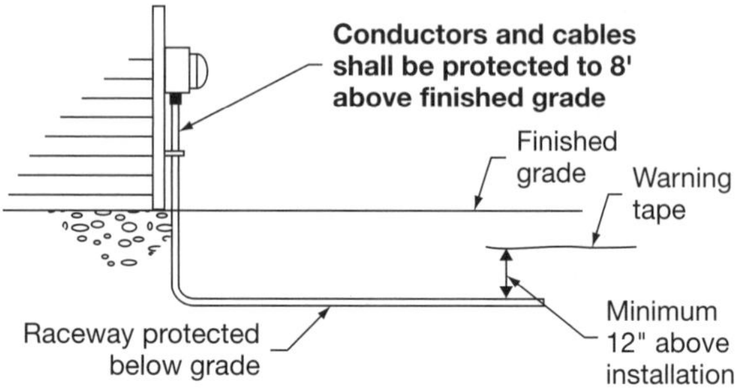

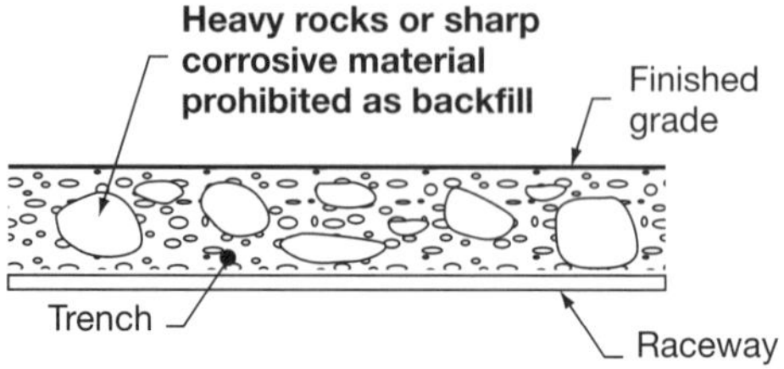

UNDERGROUND INSTALLATION REQUIREMENTS *(cont.)*

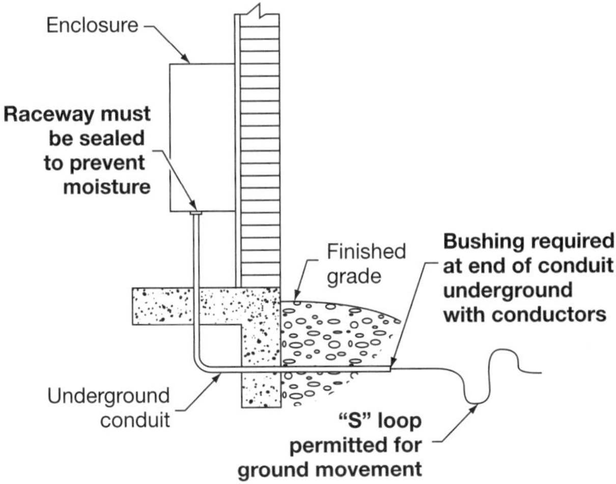

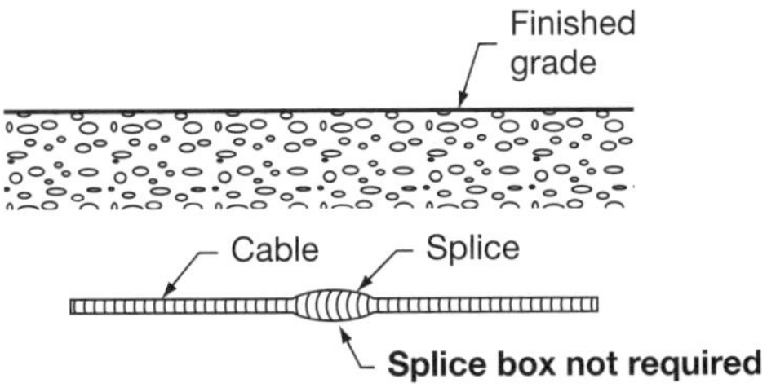

MINIMUM COVER REQUIREMENTS FOR UNDERGROUND INSTALLATIONS

Location of circuit or wiring method	Type of circuit or wiring method (0 to 600 V, nominal)				
	Direct burial cables or conductors	**Rigid metal conduit or intermediate metal conduit**	**Non-metallic raceways listed for direct burial without concrete encasement or other approved raceways**	**Residential branch circuits rated 120 V or less with GFCI protection and maximum over-current protection of 20 amperes**	**Circuits for control of irrigation and landscape lighting limited to not more than 30 V and installed with type UF or in other identified cable or raceway**
All locations not specified in these charts	24"	6"	18"	12"	6"
In trench below 2" thick concrete or equivalent	18"	6"	12"	6"	6"
Under buildings	In raceway only	—	—	In raceway only	In raceway only
Under minimum of 4" thick concrete exterior slab with no traffic and slab extending no less than 6" beyond the installation	18"	4"	4"	6" (Direct burial) 4" (In raceway)	6" (Direct burial) 4" (In raceway)
Under alleys, highways, roads, driveways, streets, and parking lots	24"	24"	24"	24"	24"

MINIMUM COVER REQUIREMENTS FOR UNDERGROUND INSTALLATIONS *(cont.)*

Location of circuit or wiring method	Type of circuit or wiring method (0 to 600 V, nominal)				
	Direct burial cables or conductors	**Rigid metal conduit or intermediate metal conduit**	**Non-metallic raceways listed for direct burial without concrete encasement or other approved raceways**	**Residential branch circuits rated 120 V or less with GFCI protection and maximum over-current protection of 20 amperes**	**Circuits for control of irrigation and landscape lighting limited to not more than 30 V and installed with type UF or in other identified cable or raceway**
Under family dwelling driveways. Outdoor parking areas used for dwelling-related purposes only	18"	18"	18"	12"	18"
In or under airport runways, including areas where the public is prohibited	18"	18"	18"	18"	18"

Notes: Cover is defined as the shortest distance between a point on the top service of any direct-buried conductor, cable, conduit or other raceway and the top surface of finished grade, concrete, etc.

Raceways approved for burial only where concrete encased shall require a 2" concrete envelope.

Lesser depths are permitted where cables/conductors rise for terminations, splices or equipment.

Where solid rock prevents compliance with the cover depths specified in this table, the wiring must be installed in raceways permitted for direct burial. The raceways shall be covered by a minimum of 2" of concrete extending down to the solid rock.

MINIMUM BURIAL DEPTH REQUIREMENTS ABOVE 600 V						
Nominal Voltage to Ground	**Directly-Buried Cables**	**Rigid Non-Metallic Conduit**	**Rigid or Intermediate Metal Conduit**	**Under 4" Minimum Building Slab**	**Cables Under or Near Airport Runways, Restricted Areas**	**Under Roads & Parking Lots**
600 to 22,000 V	30"	18"	6"	4"	18"	24"
22,000 to 40,000 V	36"	24"	6"	4"	18"	24"
Above 40,000 V	42"	30"	6"	4"	18"	24"
See NEC® Table 300.50 for complete details.						

CHAPTER 3
Ampacity, Box Fill, Branch Circuits, Conductors, Enclosures and Raceways

AMPACITY OF LAMP AND EXTENSION CORDS — TYPES S, SJ, SJT, SP, SPT AND ST			
Wire Size (AWG)	Current in Amps		
	4 Conductors	3 Conductors	2 Conductors
18	6	7	10
16	8	10	13
14	12	15	18
12	16	20	25
10	20	25	30

VERTICAL CONDUCTOR SUPPORTS		
Size of Wire (AWG, Circular Mil)	CONDUCTOR TYPE	
	Copper	Aluminum or Copper-Clad Aluminum
18 through 8	100 feet	100 feet
6 through 1/0	100 feet	200 feet
2/0 through 4/0	80 feet	180 feet
Over 4/0 through 350	60 feet	135 feet
Over 350 through 500	50 feet	120 feet
Over 500 through 750	40 feet	95 feet
Over 750	35 feet	85 feet

For SI units: one foot = 0.3048 meter.

CONDUCTOR COLOR CODE

Grounded Conductor

- White
- Gray
- Three continuous white stripes

Ungrounded Conductor

- Any color other than white, gray, or green

Equipment Grounding Conductor

- Green with one or more yellow stripes
- Bare

POWER WIRING COLOR CODE

120/240 Volt

Color	Function
Black	Phase 1
Red	Phase 2
Blue	Phase 3
Gray, White or with 3 white stripes	Neutral
Green	Ground

277/480 Volt

Color	Function
Brown	Phase 1
Orange	Phase 2
Yellow	Phase 3
Gray or with 3 white stripes	Neutral
Green with yellow stripe	Ground

POWER TRANSFORMER COLOR CODE

Wire Color	Transformer Circuit Type
Black	If a transformer does not have a tapped primary, both leads are black.
Black	If a transformer does have a tapped primary, the black is the common lead.
Black and Yellow	Tap for a tapped primary.
Black and Red	End for a tapped primary.

GROUPED CONDUCTORS

277/480 V, 3ϕ,
4-wire circuit

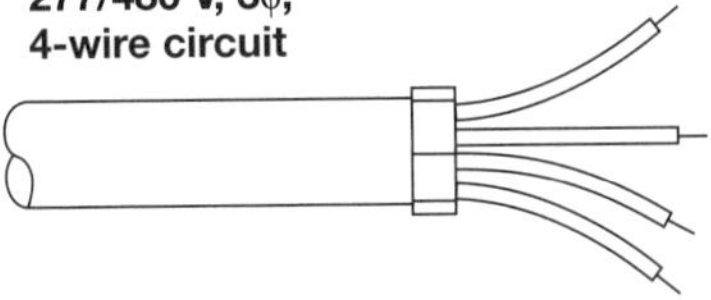

All phase conductors, grounded conductors, and equipment ground conductors shall be grouped together so as not to cause induction heating of metal raceways and enclosures.

PARALLELED CONDUCTORS

Three parallel 4/0
THHN copper conductors
per phase

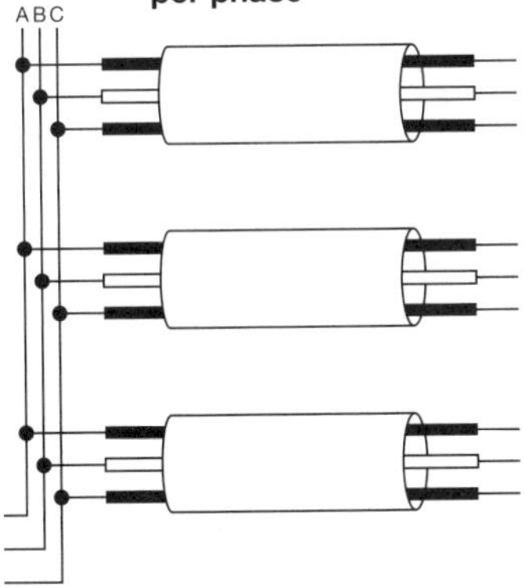

Requirements:

Same size, minimum 1/0 AWG

Same materials

Same length

Same insulation type

Same attachments

ELECTRICAL CABLE CLASS RATINGS

Electrical cable is rated according to the following parameters: the number of wires, the wire size, the type of insulation and the moisture condition of the environment of the wire. Therefore, an electrical cable designated 10/2 with ground—type UF—600 V-(UL) meets the specifications below:

- The "10" relates to wire size—10 gauge wire.
- 2 defines an electrical cable with two wires.
- The word "ground" indicates the cable has a third wire to be connected to ground.
- The term "type UF" means the insulation type has an acceptable moisture rating and is an underground feeder type cable.
- "600 V" defines the cable as being rated at 600 volts maximum.
- "(UL)" means the cable has certification from Underwriters Laboratory.

CABLE INSULATION MOISTURE RATINGS

Dry	Indoor above ground level; moisture usually not encountered.
Damp	Indoor below ground level (basement); locations are partially protected; moisture level is moderate.
Wet	Locations affected by weather (outside); concrete slabs, underground, etc.; water saturation likely.

CONDUCTOR PREFIX CODES

B	Outer braid
F	Fixture wire
FEP	Fluorinated ethylene propylene. Use in dry locations only, hotter than 90° C
H	Load temp up to 75° C
HH	Load temp up to 90° C
L	Seamless lead jacket
M	Machine tool wire
N	Resistant to oil and gas
O	Neoprene jacket
R	Rubber covering
S	Appliance cord
SP	Lamp cord, rubber
SPT	Lamp cord, plastic
T	Load temp up to 60° C
W	Wet use only
X	Moisture and heat resistant
-Z	Rated for both wet and dry use (THWN-Z)

CONDUCTOR APPLICATIONS

Type	Max. Temp	Application	Insulation	Outer Covering
FEP or FEPB	90°C (194°F) 200°C (392°F)	Dry and damp locations Dry locations – Special Apps	Fluorinated ethylene propylene	None or glass braid
MI	90°C (194°F) 250°C (482°F)	Dry and wet locations Special Apps	Magnesium oxide	Copper or alloy steel
MTW	60°C (140°F) 90°C (194°F)	Machine tool wiring – wet locations Machine tool wiring – dry locations	Flame-retardant, moisture, heat, and oil-resistant thermoplastic	None or nylon jacket
PAPER	85°C (185°F)	Underground service conductors	Paper	Lead sheath
PFA	90°C (194°F) 200°C (392°F)	Dry and damp locations Dry locations-Special Apps	Perfluoroalkoxy	None
PFAH	250°C (482°F)	Dry locations only	Perfluoroalkoxy	None
RHH	90°C (194°F)	Dry and damp locations		Moisture resistant, flame-retardant non-metallic
RHW	75°C (167°F)	Dry and wet locations	Flame-retardant, moisture-resistant thermoset	Moisture-resistant, flame-retardant, non-metallic

CONDUCTOR APPLICATIONS *(cont.)*

Type	Max. Temp	Application	Insulation	Outer Covering
RHW-2	90°C (194°F)	Dry and wet locations	Flame-retardant, moisture-resistant thermoset	Moisture-resistant, flame-retardant, non-metallic
SA	90°C (194°F) 200°C (392°F)	Dry and damp locations Special Apps	Silicone rubber	Glass or braid material
SIS	90°C (194°F)	Switchboard wiring	Flame-retardant thermostat	None
TBS	90°C (194°F)	Switchboard wiring	Thermoplastic	Flame-retardant, non-metallic
TFE	250°C (482°F)	Dry locations only	Extruded polytetrafluoroethylene	None
THHN	90°C (194°F)	Dry and damp locations	Flame-retardant, heat-resistant thermoplastic	Nylon jacket
THHW	75°C (167°F) 90°C (194°F)	Wet locations Dry locations	Flame-retardant, moisture- and heat-resistant thermoplastic	None
THW THW-2	75°C (167°F) 90°C (194°F)	Dry and wet locations Special Apps	Flame-retardant, moisture- and heat-resistant thermoplastic	None
THWN THWN-2	75°C (167°F) 90°C (194°F)	Dry and wet locations Special Apps	Flame-retardant, moisture- and heat-resistant thermoplastic	Nylon jacket
TW	60°C (140°F)	Dry and wet locations	Flame-retardant, moisture- and heat-resistant thermoplastic	None
UF	60°C (140°F) 75°C (167°F)	Refer to NEC®	Moisture-resistant Moisture- and heat-resistant	Integral with insulation

CONDUCTOR APPLICATIONS *(cont.)*

Type	Max. Temp	Application	Insulation	Outer Covering
USE USE-2	75°C (167°F) 90°C (194°F)	Refer to NEC® Dry and wet locations	Heat and moisture-resistant	Moisture-resistant non-metallic
XHH	90°C (194°F)	Dry and damp locations	Flame-retardant thermoplastic	None
XHHW	90°C (194°F) 75°C (167°F)	Dry and damp locations Wet locations	Flame-retardant, moisture-resistant thermoset	None
XHHW-2	90°C (194°F)	Dry and wet locations	Flame-retardant, moisture-resistant thermoset	None
Z	90°C (194°F) 150°C (302°F)	Dry and damp locations Dry locations – Special Apps	Modified ethylene Tetrafluoro ethylene	None
ZW	75°C (167°F) 90°C (194°F) 150°C (302°F)	Wet locations Dry and damp locations Dry locations – Special Apps	Modified ethylene Tetrafluoro ethylene	None
ZW-2	90°C (194°F)	Dry and wet locations	Modified ethylene Tetrafluoro ethylene	None

ENCLOSURE TYPES

Type	Use	Protection Against	UL Tests	Comments
1	Indoor	Incidental contact, falling dirt	Rod entry, rust resistance	—
3	Outdoor	Windblown dust, rain, sleet, and ice on enclosure	Rain, external icing, dust, and rust resistance	No protection against internal condensation, or internal icing
3R	Outdoor	Falling rain and ice on enclosure	Rod entry, rain, external icing, and rust resistance	No protection against dust, internal condensation, or internal icing
4	Indoor/outdoor	Windblown dust and rain, splashing water, hose-directed water, and ice on enclosure	Hosedown, external icing, and rust resistance	No protection against internal condensation or internal icing
4X	Indoor/outdoor	Corrosion, windblown dust and rain, splashing water, hose-directed water, and ice on enclosure	Hosedown, external icing, and corrosion resistance	No protection against internal condensation or internal icing
6	Indoor/outdoor	Occasional temporary submersion at a limited depth	—	—

ENCLOSURE TYPES *(cont.)*

Type	Use	Protection Against	UL Tests	Comments
6P	Indoor/outdoor	Prolonged submersion at a limited depth	—	—
7	Indoor locations classified as Class I, Groups A, B, C, or D, as defined in the NEC®	Withstand and contain an internal explosion of specified gases, contain an explosion sufficiently so an explosive gas–air mixture in the atmosphere is not ignited	Explosion, hydrostatic, and temperature	Enclosed heat-generating devices shall not cause external surfaces to reach temperatures capable of igniting explosive gas–air mixtures in the atmosphere
9	Indoor locations classified as Class II, Groups E, F or G, as defined in the NEC®	Dust	Dust penetration, temperature, and gasket aging	Enclosed heat-generating devices shall not cause external surfaces to reach temperatures capable of igniting explosive gas–air mixtures in the atmosphere
12	Indoor	Dust, falling dirt, and dripping noncorrosive liquids	Drip, dust, and rust resistance	No protection against internal condensation
13	Indoor	Dust, spraying water, oil and noncorrosive coolant	Oil explosion and rust resistance	No protection against internal condensation

<table>
<tr><th colspan="3">CLASSIFICATIONS OF HAZARDOUS LOCATIONS</th></tr>
<tr><td>Classes</td><td colspan="2">Likelihood that a flammable or combustible concentration is present.</td></tr>
<tr><td>I</td><td colspan="2">Sufficient quantities of flammable gases and vapors present in air to cause an explosion or ignite hazardous materials.</td></tr>
<tr><td>II</td><td colspan="2">Sufficient quantities of combustible dust are present in air to cause an explosion or ignite hazardous materials.</td></tr>
<tr><td>III</td><td colspan="2">Easily-ignitable fibers or flyings are present in air, but not in a sufficient quantity to cause an explosion or ignite hazardous materials.</td></tr>
<tr><td>Divisions</td><td colspan="2">Location containing hazardous substances.</td></tr>
<tr><td>1</td><td colspan="2">Hazardous location in which hazardous substance is normally present in air in sufficient quantities to cause an explosion or ignite hazardous materials.</td></tr>
<tr><td>2</td><td colspan="2">Hazardous location in which hazardous substance is not normally present in air in sufficient quantities to cause an explosion or ignite hazardous materials.</td></tr>
<tr><td>Groups</td><td colspan="2">Atmosphere containing flammable gases or vapors or combustible dust.</td></tr>
<tr><td>Class I</td><td>Class II</td><td>Class III</td></tr>
<tr><td>A
B
C
D</td><td>E
F
G</td><td>None</td></tr>
</table>

DIVISION 1 EXAMPLES

Class I

- Spray booth interiors.
- Areas adjacent to spraying or painting operations using volatile flammable solvents.
- Open tanks or vats of volatile flammable liquids.
- Drying or evaporation rooms for flammable solvents.
- Areas where fats and oil extraction equipment using flammable solvents are operated.
- Cleaning and dyeing plant rooms that use flammable liquids.
- Gas generator rooms.
- Pump rooms for flammable gases or volatile flammable liquids that do not contain adequate ventilation.
- Refrigeration or freezer interiors that store flammable materials.
- All other locations where sufficient ignitable quantities of flammable gases or vapors are likely to occur during routine operations.

Class II

- Grain and grain products.
- Pulverized sugar and cocoa.
- Dried egg and milk powders.
- Pulverized spices.
- Starch and pastes.
- Potato and woodflour.
- Oil meal from beans and seeds.
- Dried hay.
- Any other organic materials that may produce combustible dusts during their use or handling.

Class III

- Portions of rayon, cotton, or other textile mills.
- Manufacturing and processing plants for combustible fibers, cotton gins, and cotton seed mills.
- Flax processing plants.
- Clothing manufacturing plants.
- Woodworking plants.
- Other establishments involving similar hazardous processes or conditions.

SIZES OF PANELBOARDS

Single-Phase — 3-Wire Systems

40 A	100 A	150 A	225 A	400 A
70 A	125 A	200 A	300 A	600 A

Three-Phase — 4-Wire Systems

60 A	150 A	225 A	400 A
125 A	200 A	300 A	600 A

SIZES OF GUTTERS AND WIREWAYS

2½" × 2½" 4" × 4"	6" × 6" 8" × 8"	10" × 10"

These sizes are available in 12", 24", 36", 48", and 60" lengths.

SIZES OF DISCONNECTS

30 A 60 A 100 A	200 A 400 A 600 A	800 A 1200 A 1400 A	1600 A 1800 A

SIZES OF PULL BOXES AND JUNCTION BOXES

4" × 4" × 4"	10" × 8" × 4"	12" × 12" × 6"
6" × 4" × 4"	10" × 8" × 6"	12" × 12" × 8"
6" × 6" × 4"	10" × 10" × 4"	15" × 12" × 4"
6" × 6" × 6"	10" × 10" × 6"	15" × 12" × 6"
8" × 6" × 4"	10" × 10" × 8"	18" × 12" × 4"
8" × 6" × 6"	12" × 8" × 4"	18" × 12" × 6"
8" × 6" × 8"	12" × 8" × 6"	18" × 18" × 4"
8" × 8" × 4"	12" × 10" × 4"	18" × 18" × 6"
8" × 8" × 6"	12" × 10" × 6"	24" × 18" × 6"
8" × 8" × 8"	12" × 12" × 4"	24" × 24" × 6"
		24" × 24" × 8"

BUSWAY OR BUSDUCT		CBs AND FUSES				SWITCHBOARDS OR SWITCHGEARS	
1ϕ	3ϕ					1ϕ	3ϕ
225 A	225 A	15	70	225	800	200 A	400 A
400 A	400 A	20	80	250	1,000	400 A	600 A
600 A	600 A	25	90	300	1,200	600 A	800 A
800 A	800 A	30	100	350	1,600	800 A	1,200 A
1,000 A	1,000 A	35	110	400	2,000	1,200 A	1,600 A
1,200 A	1,200 A	40	125	450	2,500	1,600 A	2,000 A
1,350 A	1,350 A	45	150	500	3,000	2,000 A	2,500 A
1,600 A	1,600 A	50	175	600	4,000	2,500 A	3,000 A
2,000 A	2,000 A	60	200	700	5,000	3,000 A	4,000 A
2,500 A	2,500 A				6,000	4,000 A	
3,000 A	3,000 A	For fuses only, additional standard sizes are 1, 3, 6, and 10. All sizes are rated in amps.					
4,000 A	4,000 A						
5,000 A	5,000 A						

AREA DIMENSIONS OF INSULATED CONDUCTORS

Type	Size	In.^{2}
RFH-2,	18	0.0145
FFH-2	16	0.0172
RHW-2, RHH,	14	0.0293
RHW	12	0.0353
	10	0.0437
	8	0.0835
	6	0.1041
	4	0.1333
	3	0.1521
	2	0.1750
	1	0.2660
	1/0	0.3039
	2/0	0.3505
	3/0	0.4072
	4/0	0.4754
	250	0.6291
	300	0.7088
	350	0.7870
	400	0.8626
	500	1.0082
	600	1.2135
	700	1.3561
	750	1.4272
	800	1.4957
	900	1.6377
	1000	1.7719
	1250	2.3479
	1500	2.6938
	1750	3.0357
	2000	3.3719
SF-2, SFF-2	18	0.0115
	16	0.0139
	14	0.0172
SF-1, SFF-1	18	0.0065
RFH-1, XF, XFF	18	0.0080
TF, TFF, XF, XFF	16	0.0109
TW, XF, XFF, THHW, THW, THW-2	14	0.0139
TW, THHW,	12	0.0181
THW, THW-2	10	0.0243
	8	0.0437

Type	Size	In.^{2}
RHH*, RHW*, RHW-2*	14	0.0209
ABOVE AND XF, XFF	12	0.0260
RHH*, RHW*, RHW-2* XF, XFF	10	0.0333
RHH*, RHW*, RHW-2*	8	0.0556
TW, THW,	6	0.0726
THHW,	4	0.0973
THW-2,	3	0.1134
RHH*,	2	0.1333
RHW*,	1	0.1901
RHW-2*	1/0	0.2223
	2/0	0.2624
	3/0	0.3117
	4/0	0.3718
	250	0.4596
	300	0.5281
	350	0.5958
	400	0.6619
	500	0.7901
	600	0.9729
	700	1.1010
	750	1.1652
	800	1.2272
	900	1.3561
	1000	1.4784
	1250	1.8602
	1500	2.1695
	1750	2.4773
	2000	2.7818
TFN,	18	0.0055
TFFN	16	0.0072
THHN,	14	0.0097
THWN,	12	0.0133
THWN-2	10	0.0211
	8	0.0366
	6	0.0507
	4	0.0824
	3	0.0973
	2	0.1158
	1	0.1562

*Denotes types minus outer covering

AREA DIMENSIONS OF INSULATED CONDUCTORS *(cont.)*

Type	Size	In.^{2}
THHN, THWN, THWN-2 *(cont.)*	1/0	0.1855
	2/0	0.2223
	3/0	0.2679
	4/0	0.3237
	250	0.3970
	300	0.4608
	350	0.5242
	400	0.5863
	500	0.7073
	600	0.8676
	700	0.9887
	750	1.0496
	800	1.1085
	900	1.2311
	1000	1.3478
PF, PGFF, PGF, PFF, PTF, PAF, PTFF, PAFF	18	0.0058
	16	0.0075
PF, PGFF, PGF, PFF, PTF, PAF, PTFF, PAFF, TFE, FEP, PFA, FEPB, PFAH	14	0.0100
TFE, FEP, PFA, FEPB, PFAH	12	0.0137
	10	0.0191
	8	0.0333
	6	0.0468
	4	0.0670
	3	0.0804
	2	0.0973
TFE, PFAH	1	0.1399
TFE, PFA, PFAH, Z	1/0	0.1676
	2/0	0.2027
	3/0	0.2463
	4/0	0.3000
ZF, ZFF	18	0.0045
	16	0.0061
Z, ZF, ZFF	14	0.0083
Z	12	0.0117
	10	0.0191
	8	0.0302
	6	0.0430
	4	0.0625
	3	0.0855
	2	0.1029
	1	0.1269

Type	Size	In.^{2}
XHHW, ZW, XHHW-2, XHH	14	0.0139
	12	0.0181
	10	0.0243
	8	0.0437
	6	0.0590
	4	0.0814
	3	0.0962
	2	0.1146
XHHW, XHHW-2, XHH	1	0.1534
	1/0	0.1825
	2/0	0.2190
	3/0	0.2642
	4/0	0.3197
	250	0.3904
	300	0.4536
	350	0.5166
	400	0.5782
	500	0.6984
	600	0.8709
	700	0.9923
	750	1.0532
	800	1.1122
	900	1.2351
	1000	1.3519
	1250	1.7180
	1500	2.0157
	1750	2.3127
	2000	2.6073
KF-2, KFF-2	18	0.0031
	16	0.0044
	14	0.0064
	12	0.0093
	10	0.0139
KF-1, KFF-1	18	0.0026
	16	0.0037
	14	0.0055
	12	0.0083
	10	0.0127

JUNCTION BOX CALCULATIONS

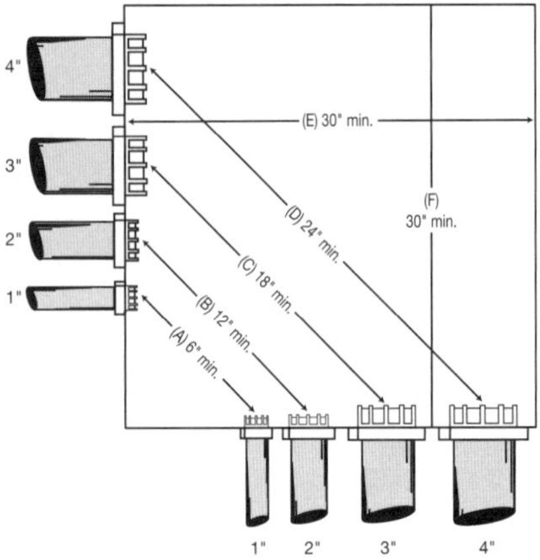

Distance (A) is 6 × 1" = 6" minimum
Distance (B) is 6 × 2" = 12" minimum
Distance (C) is 6 × 3" = 18" minimum
Distance (D) is 6 × 4" = 24" minimum
Distance (E) is (6 × 4") + 3" + 2" + 1" = 30" minimum
Distance (F) is (6 × 4") + 3" + 2" + 1" = 30" minimum

CONDUCTOR VOLUME ALLOWANCE

Wire Size (AWG)	Volume Each (In.^{3})	Formula
18	1.50	V = L · W · D
16	1.75	Volume =
14	2.00	Length times width
12	2.25	times depth =
10	2.50	cubic inches
8	3.00	
6	5.00	

To find box size needed, add up total volume for all wires to be used. Then use the volume formula; e.g., if total volume of all wires is 420 cu. in. – use an 8" × 10" × 6" box = 480 cu/in.

BOX FILL

Max. Number of Conductors in Outlet, Device and Junction Boxes

Box Dimension in Inches Trade Size or Type	Min. Capacity (In.³)	Maximum Number of Conductors per Wire Size (AWG)						
		# 18	# 16	# 14	# 12	# 10	# 8	# 6
4 × 1¼ round or octagonal	12.5	8	7	6	5	5	5	2
4 × 1½ round or octagonal	15.5	10	8	7	6	6	5	3
4 × 2⅛ round or octagonal	21.5	14	12	10	9	8	7	4
4 × 1¼ square	18.0	12	10	9	8	7	6	3
4 × 1½ square	21.0	14	12	10	9	8	7	4
4 × 2⅛ square	30.3	20	17	15	13	12	10	6
4¹¹⁄₁₆ × 1¼ square	25.5	17	14	12	11	10	8	5
4¹¹⁄₁₆ × 1½ square	29.5	19	16	14	13	11	9	5
4¹¹⁄₁₆ × 2⅛ square	42.0	28	24	21	18	16	14	8
3 × 2 × 1½ device	7.5	5	4	3	3	3	2	1
3 × 2 × 2 device	10.0	6	5	5	4	4	3	2
3 × 2 × 2¼ device	10.5	7	6	5	4	4	3	2
3 × 2 × 2½ device	12.5	8	7	6	5	5	4	2
3 × 2 × 2¾ device	14.0	9	8	7	6	5	4	2
3 × 2 × 3½ device	18.0	12	10	9	8	7	6	3
4 × 2⅛ × 1½ device	10.3	6	5	5	4	4	3	2
4 × 2⅛ × 1⅞ device	13.0	8	7	6	5	5	4	2
4 × 2⅛ × 2⅛ device	14.5	9	8	7	6	5	4	2
3¾ × 2 × 2½ masonry box/gang	14.0	9	8	7	6	5	4	2
3¾ × 2 × 3½ masonry box/gang	21.0	14	12	10	9	8	7	2

Where one or more internal cable clamps are present in the box, a single volume allowance (conductor) shall be made based on the largest size conductor in the box.

RIGID METAL CONDUIT—MAXIMUM NUMBER OF CONDUCTORS

Type	Size	RMC Trade Size (In.)											
		1/2	3/4	1	1 1/4	1 1/2	2	2 1/2	3	3 1/2	4	5	6
RHH, RHW,	14	4	7	12	21	28	46	66	102	136	176	276	398
RHW-2	12	3	6	10	17	23	38	55	85	113	146	229	330
	10	3	5	8	14	19	31	44	68	91	118	185	267
	8	1	2	4	7	10	16	23	36	48	61	97	139
	6	1	1	3	6	8	13	18	29	38	49	77	112
	4	1	1	2	4	6	10	14	22	30	38	60	87
	3	1	1	2	4	5	9	12	19	26	34	53	76
	2	1	1	1	3	4	7	11	17	23	29	46	66
	1	0	1	1	1	3	5	7	11	15	19	30	44
	1/0	0	1	1	1	2	4	6	10	13	17	26	38
	2/0	0	1	1	1	2	4	5	8	11	14	23	33
	3/0	0	0	1	1	1	3	4	7	10	12	20	28
	4/0	0	0	1	1	1	3	4	6	8	11	17	24
	250	0	0	0	1	1	1	3	4	6	8	13	18
	300	0	0	0	1	1	1	2	4	5	7	11	16
	350	0	0	0	1	1	1	2	4	5	6	10	15
	400	0	0	0	1	1	1	1	3	4	6	9	13
	500	0	0	0	1	1	1	1	3	4	5	8	11
	600	0	0	0	0	1	1	1	2	3	4	6	9
	700	0	0	0	0	1	1	1	1	3	4	6	8
	750	0	0	0	0	0	1	1	1	3	3	5	8
	800	0	0	0	0	0	1	1	1	2	3	5	7
	1,000	0	0	0	0	0	1	1	1	1	3	4	6
TW	14	9	15	25	44	59	98	140	216	288	370	581	839
	12	7	12	19	33	45	75	107	165	221	284	446	644
	10	5	9	14	25	34	56	80	123	164	212	332	480
	8	3	5	8	14	19	31	44	68	91	118	185	267
RHH*, RHW*,	14	6	10	17	29	39	65	93	143	191	246	387	558
RHW-2*, THHW,	12	5	8	13	23	32	52	75	115	154	198	311	448
THW, THW-2	10	3	6	10	18	25	41	58	90	120	154	242	350
	8	1	4	6	11	15	24	35	54	72	92	145	209
RHH*, RHW*,	6	1	3	5	8	11	18	27	41	55	71	111	160
RHW-2*, TW,	4	1	1	3	6	8	14	20	31	41	53	83	120
THW, THHW,	3	1	1	3	5	7	12	17	26	35	45	71	103
THW-2	2	1	1	2	4	6	10	14	22	30	38	60	87
	1	1	1	1	3	4	7	10	15	21	27	42	61
	1/0	0	1	1	2	3	6	8	13	18	23	36	52
	2/0	0	1	1	2	3	5	7	11	15	19	31	44
NOTE:	3/0	0	1	1	1	2	4	6	9	13	16	26	37
(*) Denotes	4/0	0	0	1	1	1	3	5	8	10	14	21	31
Types Minus	250	0	0	1	1	1	3	4	6	8	11	17	25
Outer	300	0	0	1	1	1	2	3	5	7	9	15	22
Covering	350	0	0	0	1	1	1	3	5	6	8	13	19
	400	0	0	0	1	1	1	3	4	6	7	12	17
	500	0	0	0	1	1	1	2	3	5	6	10	14
	600	0	0	0	1	1	1	1	3	4	5	8	12
	700	0	0	0	0	1	1	1	2	3	4	7	10
	750	0	0	0	0	1	1	1	2	3	4	7	10
	800	0	0	0	0	1	1	1	2	3	4	6	9
	1,000	0	0	0	0	0	1	1	1	2	3	5	8
THHN,	14	13	22	36	63	85	140	200	309	412	531	833	1,202
THWN,	12	9	16	26	46	62	102	146	225	301	387	608	877
THWN-2	10	6	10	17	29	39	64	92	142	189	244	383	552
	8	3	6	9	16	22	37	53	82	109	140	221	318
	6	2	4	7	12	16	27	38	59	79	101	159	230
	4	1	2	4	7	10	16	23	36	48	62	98	141
	3	1	1	3	6	8	14	20	31	41	53	83	120
	2	1	1	3	5	7	11	17	26	34	44	70	100
	1	1	1	1	4	5	8	12	19	25	33	51	74

RIGID METAL CONDUIT—MAXIMUM NUMBER OF CONDUCTORS *(cont.)*

Type	Size	RMC Trade Size (In.) ½	¾	1	1¼	1½	2	2½	3	3½	4	5	6
THHN, THWN, THWN-2 *(cont.)*	1/0	1	1	1	3	4	7	10	16	21	27	43	63
	2/0	0	1	1	2	3	6	8	13	18	23	36	52
	3/0	0	1	1	1	3	5	7	11	15	19	30	43
	4/0	0	1	1	1	2	4	6	9	12	16	25	36
	250	0	0	1	1	1	3	5	7	10	13	20	29
	300	0	0	1	1	1	3	4	6	8	11	17	25
	350	0	0	1	1	1	2	3	5	7	10	15	22
	400	0	0	1	1	1	2	3	5	7	8	13	20
	500	0	0	0	1	1	1	2	4	5	7	11	16
	600	0	0	0	1	1	1	1	3	4	6	9	13
	700	0	0	0	1	1	1	1	3	4	5	8	11
	750	0	0	0	0	1	1	1	3	4	5	7	11
	800	0	0	0	0	1	1	1	2	3	4	7	10
	1,000	0	0	0	0	1	1	1	1	3	4	6	8
FEP, FEPB, PFA, PFAH, TFE	14	12	22	35	61	83	136	194	300	400	515	808	1,166
	12	9	16	26	44	60	99	142	219	292	376	590	851
	10	6	11	18	32	43	71	102	157	209	269	423	610
	8	3	6	10	18	25	41	58	90	120	154	242	350
	6	2	4	7	13	17	29	41	64	85	110	172	249
	4	1	3	5	9	12	20	29	44	59	77	120	174
	3	1	2	4	7	10	17	24	37	50	64	100	145
	2	1	1	3	6	8	14	20	31	41	53	83	120
PFA, PFAH, TFE	1	1	1	2	4	6	9	14	21	28	37	57	83
PFA, PFAH, TFE, Z	1/0	1	1	1	3	5	8	11	18	24	30	48	69
	2/0	1	1	1	3	4	6	9	14	19	25	40	57
	3/0	0	1	1	2	3	5	8	12	16	21	33	47
	4/0	0	1	1	1	2	4	6	10	13	17	27	39
Z	14	15	26	42	73	100	164	234	361	482	621	974	1,405
	12	10	18	30	52	71	116	166	256	342	440	691	997
	10	6	11	18	32	43	71	102	157	209	269	423	610
	8	4	7	11	20	27	45	64	99	132	170	267	386
	6	3	5	8	14	19	31	45	69	93	120	188	271
	4	1	3	5	9	13	22	31	48	64	82	129	186
	3	1	2	4	7	9	16	22	35	47	60	94	136
	2	1	1	3	6	8	13	19	29	39	50	78	113
	1	1	1	2	5	6	10	15	23	31	40	63	92
XHH, XHHW, XHHW-2, ZW	14	9	15	25	44	59	98	140	216	288	370	581	839
	12	7	12	19	33	45	75	107	165	221	284	446	644
	10	5	9	14	25	34	56	80	123	164	212	332	480
	8	3	5	8	14	19	31	44	68	91	118	185	267
	6	1	3	6	10	14	23	33	51	68	87	137	197
	4	1	2	4	7	10	16	24	37	49	63	99	143
	3	1	1	3	6	8	14	20	31	41	53	84	121
	2	1	1	3	5	7	12	17	26	35	45	70	101
	1	1	1	1	4	5	9	12	19	26	33	52	76
XHH, XHHW, XHHW-2	1/0	1	1	1	3	4	7	10	16	22	28	44	64
	2/0	0	1	1	2	3	6	9	13	18	23	37	53
	3/0	0	1	1	1	3	5	7	11	15	19	30	44
	4/0	0	1	1	1	2	4	6	9	12	16	25	36
	250	0	0	1	1	1	3	5	7	10	13	20	30
	300	0	0	1	1	1	3	4	6	9	11	18	25
	350	0	0	1	1	1	2	3	6	7	10	15	22
	400	0	0	1	1	1	2	3	5	7	9	14	20
	500	0	0	0	1	1	1	2	4	5	7	11	16
	600	0	0	0	1	1	1	1	3	4	6	9	13
	700	0	0	0	1	1	1	1	3	4	5	8	11
	750	0	0	0	0	1	1	1	3	4	5	7	11
	800	0	0	0	0	1	1	1	2	3	4	7	10

LIQUIDTIGHT FLEXIBLE METAL CONDUIT—MAXIMUM NUMBER OF CONDUCTORS

Type	Size	LFMC Trade Size (In.) ½	¾	1	1¼	1½	2	2½	3	3½	4
RHH, RHW,	14	4	7	12	21	27	44	66	102	133	173
RHW-2	12	3	6	10	17	22	36	55	84	110	144
	10	3	5	8	14	18	29	44	68	89	116
	8	1	2	4	7	9	15	23	36	46	61
	6	1	1	3	6	7	12	18	28	37	48
	4	1	1	2	4	6	9	14	22	29	38
	3	1	1	1	4	5	8	13	19	25	33
	2	1	1	1	3	4	7	11	17	22	29
	1	0	1	1	1	3	5	7	11	14	19
	1/0	0	1	1	1	2	4	6	10	13	16
	2/0	0	1	1	1	1	3	5	8	11	14
	3/0	0	0	1	1	1	3	4	7	9	12
	4/0	0	0	1	1	1	2	4	6	8	10
	250	0	0	0	1	1	1	3	4	6	8
	300	0	0	0	1	1	1	2	4	5	7
	350	0	0	0	1	1	1	2	3	5	6
	400	0	0	0	1	1	1	1	3	4	6
	500	0	0	0	1	1	1	1	3	4	5
	600	0	0	0	0	1	1	1	2	3	4
	700	0	0	0	0	0	1	1	1	3	3
	750	0	0	0	0	0	1	1	1	2	3
	800	0	0	0	0	0	1	1	1	2	3
	1,000	0	0	0	0	0	1	1	1	1	3
TW	14	9	15	25	44	57	93	140	215	280	365
	12	7	12	19	33	43	71	108	165	215	280
	10	5	9	14	25	32	53	80	123	160	209
	8	3	5	8	14	18	29	44	68	89	116
RHH*, RHW*,	14	6	10	16	29	38	62	93	143	186	243
RHW-2*, THHW,	12	5	8	13	23	30	50	75	115	149	195
THW, THW-2	10	3	6	10	18	23	39	58	89	117	152
	8	1	4	6	11	14	23	35	53	70	91
RHH*, RHW*,	6	1	3	5	8	11	18	27	41	53	70
RHW-2*, TW,	4	1	1	3	6	8	13	20	30	40	52
THW, THHW,	3	1	1	3	5	7	11	17	26	34	44
THW-2	2	1	1	2	4	6	9	14	22	29	38
	1	1	1	1	3	4	7	10	15	20	26
	1/0	0	1	1	2	3	6	8	13	17	23
	2/0	0	1	1	2	3	5	7	11	15	19
Note:	3/0	0	1	1	1	2	4	6	9	12	16
(*) Denotes	4/0	0	0	1	1	1	3	5	8	10	13
types minus	250	0	0	1	1	1	3	4	6	8	11
outer	300	0	0	1	1	1	2	3	5	7	9
covering	350	0	0	0	1	1	1	3	5	6	8
	400	0	0	0	1	1	1	3	4	6	7
	500	0	0	0	1	1	1	2	3	5	6
	600	0	0	0	1	1	1	1	3	4	5
	700	0	0	0	0	1	1	1	2	3	4
	750	0	0	0	0	1	1	1	2	3	4
	800	0	0	0	0	1	1	1	2	3	4
	1,000	0	0	0	0	0	1	1	1	2	3
THHN,	14	13	22	36	63	81	133	201	308	401	523
THWN,	12	9	16	26	46	59	97	146	225	292	381
THWN-2	10	6	10	16	29	37	61	92	141	184	240
	8	3	6	9	16	21	35	53	81	106	138
	6	2	4	7	12	15	25	38	59	76	100
	4	1	2	4	7	9	15	23	36	47	61
	3	1	1	3	6	8	13	20	30	40	52
	2	1	1	3	5	7	11	17	26	33	44
	1	1	1	1	4	5	8	12	19	25	32

LIQUIDTIGHT FLEXIBLE METAL CONDUIT—MAXIMUM NUMBER OF CONDUCTORS *(cont.)*

		LFMC Trade Size (In.)									
Type	**Size**	**½**	**¾**	**1**	**1¼**	**1½**	**2**	**2½**	**3**	**3½**	**4**
THHN, THWN, THWN-2 *(cont.)*	1/0	1	1	1	3	4	7	10	16	21	27
	2/0	0	1	1	2	3	6	8	13	17	23
	3/0	0	1	1	1	3	5	7	11	14	19
	4/0	0	1	1	1	2	4	6	9	12	15
	250	0	0	1	1	1	3	5	7	10	12
	300	0	0	1	1	1	3	4	6	8	11
	350	0	0	1	1	1	2	3	5	7	9
	400	0	0	0	1	1	1	3	5	6	8
	500	0	0	0	1	1	1	2	4	5	7
	600	0	0	0	1	1	1	1	3	4	6
	700	0	0	0	1	1	1	1	3	4	5
	750	0	0	0	0	1	1	1	3	3	5
	800	0	0	0	0	1	1	1	2	3	4
	1,000	0	0	0	0	0	1	1	1	3	3
FEP, FEPB, PFA, PFAH, TFE	14	12	21	35	61	79	129	195	299	389	507
	12	9	15	25	44	57	94	142	218	284	370
	10	6	11	18	32	41	68	102	156	203	266
	8	3	6	10	18	23	39	58	89	117	152
	6	2	4	7	13	17	27	41	64	83	108
	4	1	3	5	9	12	19	29	44	58	75
	3	1	2	4	7	10	16	24	37	48	63
	2	1	1	3	6	8	13	20	30	40	52
PFA, PFAH, TFE	1	1	1	2	4	5	9	14	21	28	36
PFA, PFAH, TFE, Z	1/0	1	1	1	3	4	7	11	18	23	30
	2/0	1	1	1	3	4	6	9	14	19	25
	3/0	0	1	1	2	3	5	8	12	16	20
	4/0	0	1	1	1	2	4	6	10	13	17
Z	14	20	26	42	73	95	156	235	360	469	611
	12	14	18	30	52	67	111	167	255	332	434
	10	8	11	18	32	41	68	102	156	203	266
	8	5	7	11	20	26	43	64	99	129	168
	6	4	5	8	14	18	30	45	69	90	118
	4	2	3	5	9	12	20	31	48	62	81
	3	2	2	4	7	9	15	23	35	45	59
	2	1	1	3	6	7	12	19	29	38	49
	1	1	1	2	5	6	10	15	23	30	40
XHH, XHHW, XHHW-2, ZW	14	9	15	25	44	57	93	140	215	280	365
	12	7	12	19	33	43	71	108	165	215	280
	10	5	9	14	25	32	53	80	123	160	209
	8	3	5	8	14	18	29	44	68	89	116
	6	1	3	6	10	13	22	33	50	66	86
	4	1	2	4	7	9	16	24	36	48	62
	3	1	1	3	6	8	13	20	31	40	52
	2	1	1	3	5	7	11	17	26	34	44
	1	1	1	1	4	5	8	12	19	25	33
XHH, XHHW, XHHW-2	1/0	1	1	1	3	4	7	10	16	21	28
	2/0	0	1	1	2	3	6	9	13	17	23
	3/0	0	1	1	1	3	5	7	11	14	19
	4/0	0	1	1	1	2	4	6	9	12	16
	250	0	0	1	1	1	3	5	7	10	13
	300	0	0	1	1	1	3	4	6	8	11
	350	0	0	1	1	1	2	3	5	7	10
	400	0	0	0	1	1	1	3	5	6	8
	500	0	0	0	1	1	1	2	4	5	7
	600	0	0	0	1	1	1	1	3	4	6
	700	0	0	0	1	1	1	1	3	4	5
	750	0	0	0	0	1	1	1	3	3	5
	800	0	0	0	0	1	1	1	2	3	4

NONMETALLIC TUBING—MAXIMUM NUMBER OF CONDUCTORS

Type	Size	ENT Trade Size (In.) ½	¾	1	1¼	1½	2
RHH, RHW, RHW-2	14	3	6	10	19	26	43
	12	2	5	9	16	22	36
	10	1	4	7	13	17	29
	8	1	1	3	6	9	15
	6	1	1	3	5	7	12
	4	1	1	2	4	6	9
	3	1	1	1	3	5	8
	2	0	1	1	3	4	7
	1	0	1	1	1	3	5
	1/0	0	0	1	1	2	4
	2/0	0	0	1	1	1	3
	3/0	0	0	1	1	1	3
	4/0	0	0	1	1	1	2
	250	0	0	0	1	1	1
	300	0	0	0	1	1	1
	350	0	0	0	1	1	1
	400	0	0	0	1	1	1
	500	0	0	0	0	1	1
	600	0	0	0	0	1	1
	700	0	0	0	0	0	1
	750	0	0	0	0	0	1
	800	0	0	0	0	0	1
	1,000	0	0	0	0	0	1
TW	14	7	13	22	40	55	92
	12	5	10	17	31	42	71
	10	4	7	13	23	32	52
	8	1	4	7	13	17	29
RHH*, RHW*, RHW-2*, THHW, THW,THW-2	14	4	8	15	27	37	61
	12	3	7	12	21	29	49
	10	3	5	9	17	23	38
	8	1	3	5	10	14	23
RHH*, RHW*, RHW-2*, TW, THW, THHW, THW-2	6	1	2	4	7	10	17
	4	1	1	3	5	8	13
	3	1	1	2	5	7	11
	2	1	1	2	4	6	9
	1	0	1	1	3	4	6
	1/0	0	1	1	2	3	5
	2/0	0	1	1	1	3	5
NOTE:	3/0	0	0	1	1	2	4
(*) Denotes	4/0	0	0	1	1	1	3
types minus	250	0	0	1	1	1	2
outer	300	0	0	0	1	1	2
covering	350	0	0	0	1	1	1
	400	0	0	0	1	1	1
	500	0	0	0	1	1	1
	600	0	0	0	0	1	1
	700	0	0	0	0	1	1
	750	0	0	0	0	1	1
	800	0	0	0	0	1	1
	1,000	0	0	0	0	0	1
THHN, THWN, THWN-2	14	10	18	32	58.	80	132
	12	7	13	23	42	58	96
	10	4	8	15	26	36	60
	8	2	5	8	15	21	35
	6	1	3	6	11	15	25
	4	1	1	4	7	9	15
	3	1	1	3	5	8	13
	2	1	1	2	5	6	11
	1	1	1	1	3	5	8

NONMETALLIC TUBING—MAXIMUM NUMBER OF CONDUCTORS *(cont.)*

		ENT Trade Size (In.)					
Type	**Size**	**½**	**¾**	**1**	**1¼**	**1½**	**2**
THHN,	1/0	0	1	1	3	4	7
THWN,	2/0	0	1	1	2	3	5
THWN-2	3/0	0	1	1	1	3	4
(cont.)	4/0	0	0	1	1	2	4
	250	0	0	1	1	1	3
	300	0	0	1	1	1	2
	350	0	0	0	1	1	2
	400	0	0	0	1	1	1
	500	0	0	0	1	1	1
	600	0	0	0	1	1	1
	700	0	0	0	0	1	1
	750	0	0	0	0	1	1
	800	0	0	0	0	1	1
	1,000	0	0	0	0	0	1
FEP, FEPB,	14	10	18	31	56	77	128
PFA, PFAH,	12	7	13	23	41	56	93
TFE	10	5	9	16	29	40	67
	8	3	5	9	17	23	38
	6	1	4	6	12	16	27
	4	1	2	4	8	11	19
	3	1	1	4	7	9	16
	2	1	1	3	5	8	13
PFA, PFAH, TFE	1	1	1	1	4	5	9
PFA, PFAH,	1/0	0	1	1	3	4	7
TFE, Z	2/0	0	1	1	2	4	6
	3/0	0	1	1	1	3	5
	4/0	0	1	1	1	2	4
Z	14	12	22	38	68	93	154
	12	8	15	27	48	66	109
	10	5	9	16	29	40	67
	8	3	6	10	18	25	42
	6	1	4	7	13	18	30
	4	1	3	5	9	12	20
	3	1	1	3	6	9	15
	2	1	1	3	5	7	12
	1	1	1	2	4	6	10
XHH,	14	7	13	22	40	55	92
XHHW,	12	5	10	17	31	42	71
XHHW-2,	10	4	7	13	23	32	52
ZW	8	1	4	7	13	17	29
	6	1	3	5	9	13	21
	4	1	1	4	7	9	15
	3	1	1	3	6	8	13
	2	1	1	2	5	6	11
XHH,	1	1	1	1	3	5	8
XHHW,	1/0	0	1	1	3	4	7
XHHW-2	2/0	0	1	1	2	3	6
	3/0	0	1	1	1	3	5
	4/0	0	0	1	1	2	4
	250	0	0	1	1	1	3
	300	0	0	1	1	1	3
	350	0	0	1	1	1	2
	400	0	0	0	1	1	1
	500	0	0	0	1	1	1
	600	0	0	0	1	1	1
	700	0	0	0	0	1	1
	750	0	0	0	0	1	1
	800	0	0	0	0	1	1

ELECTRICAL METALLIC TUBING—MAXIMUM NUMBER OF CONDUCTORS

		EMT Trade Size (In.)									
Type	**Size**	**½**	**¾**	**1**	**1¼**	**1½**	**2**	**2½**	**3**	**3½**	**4**
RHH, RHW, RHW-2	14	4	7	11	20	27	46	80	120	157	201
	12	3	6	9	17	23	38	66	100	131	167
	10	2	5	8	13	18	30	53	81	105	135
	8	1	2	4	7	9	16	28	42	55	70
	6	1	1	3	5	8	13	22	34	44	56
	4	1	1	2	4	6	10	17	26	34	44
	3	1	1	1	4	5	9	15	23	30	38
	2	1	1	1	3	4	7	13	20	26	33
	1	0	1	1	1	3	5	9	13	17	22
	1/0	0	1	1	1	2	4	7	11	15	19
	2/0	0	1	1	1	2	4	6	10	13	17
	3/0	0	0	1	1	1	3	5	8	11	14
	4/0	0	0	1	1	1	3	5	7	9	12
	250	0	0	0	1	1	1	3	5	7	9
	300	0	0	0	1	1	1	3	5	6	8
	350	0	0	0	1	1	1	3	4	6	7
	400	0	0	0	1	1	1	2	4	5	7
	500	0	0	0	0	1	1	2	3	4	6
	600	0	0	0	0	1	1	1	3	4	5
	700	0	0	0	0	0	1	1	2	3	4
	750	0	0	0	0	0	1	1	2	3	4
	800	0	0	0	0	0	1	1	2	3	4
	1,000	0	0	0	0	0	1	1	1	2	3
TW	14	8	15	25	43	58	96	168	254	332	424
	12	6	11	19	33	45	74	129	195	255	326
	10	5	8	14	24	33	55	96	145	190	243
	8	2	5	8	13	18	30	53	81	105	135
RHH*, RHW*, RHW-2*, THHW, THW, THW-2	14	6	10	16	28	39	64	112	169	221	282
	12	4	8	13	23	31	51	90	136	177	227
	10	3	6	10	18	24	40	70	106	138	177
	8	1	4	6	10	14	24	42	63	83	106
RHH*, RHW*, RHW-2*, TW, THW, THHW, THW-2	6	1	3	4	8	11	18	32	48	63	81
	4	1	1	3	6	8	13	24	36	47	60
	3	1	1	3	5	7	12	20	31	40	52
	2	1	1	2	4	6	10	17	26	34	44
	1	1	1	1	3	4	7	12	18	24	31
	1/0	0	1	1	2	3	6	10	16	20	26
	2/0	0	1	1	1	3	5	9	13	17	22
NOTE:	3/0	0	1	1	1	2	4	7	11	15	19
(*) Denotes types minus outer covering	4/0	0	0	1	1	1	3	6	9	12	16
	250	0	0	1	1	1	3	5	7	10	13
	300	0	0	1	1	1	2	4	6	8	11
	350	0	0	0	1	1	1	4	6	7	10
	400	0	0	0	1	1	1	3	5	7	9
	500	0	0	0	1	1	1	3	4	6	7
	600	0	0	0	1	1	1	2	3	4	6
	700	0	0	0	0	1	1	1	3	4	5
	750	0	0	0	0	1	1	1	3	4	5
	800	0	0	0	0	1	1	1	3	3	5
	1,000	0	0	0	0	0	1	1	2	3	4
THHN, THWN, THWN-2	14	12	22	35	61	84	138	241	364	476	608
	12	9	16	26	45	61	101	176	266	347	443
	10	5	10	16	28	38	63	111	167	219	279
	8	3	6	9	16	22	36	64	96	126	161
	6	2	4	7	12	16	26	46	69	91	116
	4	1	2	4	7	10	16	28	43	56	71
	3	1	1	3	6	8	13	24	36	47	60
	2	1	1	3	5	7	11	20	30	40	51
	1	1	1	1	4	5	8	15	22	29	37

ELECTRICAL METALLIC TUBING—MAXIMUM NUMBER OF CONDUCTORS *(cont.)*

		EMT Trade Size (In.)									
Type	**Size**	**½**	**¾**	**1**	**1¼**	**1½**	**2**	**2½**	**3**	**3½**	**4**
THHN,	1/0	1	1	1	3	4	7	12	19	25	32
THWN,	2/0	0	1	1	2	3	6	10	16	20	26
THWN-2	3/0	0	1	1	1	3	5	8	13	17	22
(cont.)	4/0	0	1	1	1	2	4	7	11	14	18
	250	0	0	1	1	1	3	6	9	11	15
	300	0	0	1	1	1	3	5	7	10	13
	350	0	0	1	1	1	2	4	6	9	11
	400	0	0	0	1	1	1	4	6	8	10
	500	0	0	0	1	1	1	3	5	6	8
	600	0	0	0	1	1	1	2	4	5	7
	700	0	0	0	1	1	1	2	3	4	6
	750	0	0	0	0	1	1	1	3	4	5
	800	0	0	0	0	1	1	1	3	4	5
	1,000	0	0	0	0	1	1	1	2	3	4
FEP, FEPB,	14	12	21	34	60	81	134	234	354	462	590
PFA, PFAH,	12	9	15	25	43	59	98	171	258	337	430
TFE	10	6	11	18	31	42	70	122	185	241	309
	8	3	6	10	18	24	40	70	106	138	177
	6	2	4	7	12	17	28	50	75	98	126
	4	1	3	5	9	12	20	35	53	69	88
	3	1	2	4	7	10	16	29	44	57	73
	2	1	1	3	6	8	13	24	36	47	60
PFA, PFAH, TFE	1	1	1	2	4	6	9	16	25	33	42
PFA, PFAH,	1/0	1	1	1	3	5	8	14	21	27	35
TFE, Z	2/0	0	1	1	3	4	6	11	17	22	29
	3/0	0	1	1	2	3	5	9	14	18	24
	4/0	0	1	1	1	2	4	8	11	15	19
Z	14	14	25	41	72	98	161	282	426	556	711
	12	10	18	29	51	69	114	200	302	394	504
	10	6	11	18	31	42	70	122	185	241	309
	8	4	7	11	20	27	44	77	117	153	195
	6	3	5	8	14	19	31	54	82	107	137
	4	1	3	5	9	13	21	37	56	74	94
	3	1	2	4	7	9	15	27	41	54	69
	2	1	1	3	6	8	13	22	34	45	57
	1	1	1	2	4	6	10	18	28	36	46
XHH,	14	8	15	25	43	58	96	168	254	332	424
XHHW,	12	6	11	19	33	45	74	129	195	255	326
XHHW-2,	10	5	8	14	24	33	55	96	145	190	243
ZW	8	2	5	8	13	18	30	53	81	105	135
	6	1	3	6	10	14	22	39	60	78	100
	4	1	2	4	7	10	16	28	43	56	72
	3	1	1	3	6	8	14	24	36	48	61
	2	1	1	3	5	7	11	20	31	40	51
XHH,	1	1	1	1	4	5	8	15	23	30	38
XHHW,	1/0	1	1	1	3	4	7	13	19	25	32
XHHW-2	2/0	0	1	1	2	3	6	10	16	21	27
	3/0	0	1	1	1	3	5	9	13	17	22
	4/0	0	1	1	1	2	4	7	11	14	18
	250	0	0	1	1	1	3	6	9	12	15
	300	0	0	1	1	1	3	5	8	10	13
	350	0	0	1	1	1	2	4	7	9	11
	400	0	0	0	1	1	1	4	6	8	10
	500	0	0	0	1	1	1	3	5	6	8
	600	0	0	0	1	1	1	2	4	5	6
	700	0	0	0	0	1	1	2	3	4	6
	750	0	0	0	0	1	1	1	3	4	5
	800	0	0	0	0	1	1	1	3	4	5

FLEXIBLE METAL CONDUIT—MAXIMUM NUMBER OF CONDUCTORS

Type	Size	FMC Trade Size (In.) 1/2	3/4	1	1 1/4	1 1/2	2	2 1/2	3	3 1/2	4
RHH, RHW,	14	4	7	11	17	25	44	67	96	131	171
RHW-2	12	3	6	9	14	21	37	55	80	109	142
	10	3	5	7	11	17	30	45	64	88	115
	8	1	2	4	6	9	15	23	34	46	60
	6	1	1	3	5	7	12	19	27	37	48
	4	1	1	2	4	5	10	14	21	29	37
	3	1	1	1	3	5	8	13	18	25	33
	2	1	1	1	3	4	7	11	16	22	28
	1	0	1	1	1	2	5	7	10	14	19
	1/0	0	1	1	1	2	4	6	9	12	16
	2/0	0	1	1	1	1	3	5	8	11	14
	3/0	0	0	1	1	1	3	5	7	9	12
	4/0	0	0	1	1	1	2	4	6	8	10
	250	0	0	0	1	1	1	3	4	6	8
	300	0	0	0	1	1	1	2	4	5	7
	350	0	0	0	1	1	1	2	3	5	6
	400	0	0	0	0	1	1	1	3	4	6
	500	0	0	0	0	1	1	1	3	4	5
	600	0	0	0	0	1	1	1	2	3	4
	700	0	0	0	0	0	1	1	1	3	3
	750	0	0	0	0	0	1	1	1	2	3
	800	0	0	0	0	0	1	1	1	2	3
	1,000	0	0	0	0	0	1	1	1	1	3
TW	14	9	15	23	36	53	94	141	203	277	361
	12	7	11	18	28	41	72	108	156	212	277
	10	5	8	13	21	30	54	81	116	158	207
	8	3	5	7	11	17	30	45	64	88	115
RHH*, RHW*,	14	6	10	15	24	35	62	94	135	184	240
RHW-2*, THHW,	12	5	8	12	19	28	50	75	108	148	193
THW, THW-2	10	4	6	10	15	22	39	59	85	115	151
	8	1	4	6	9	13	23	35	51	69	90
RHH*, RHW*,	6	1	3	4	7	10	18	27	39	53	69
RHW-2*, TW,	4	1	1	3	5	7	13	20	29	39	51
THW, THHW,	3	1	1	3	4	6	11	17	25	34	44
THW-2	2	1	1	2	4	5	10	14	21	29	37
	1	1	1	1	2	4	7	10	15	20	26
	1/0	0	1	1	1	3	6	9	12	17	22
	2/0	0	1	1	1	3	5	7	10	14	19
Note:	3/0	0	1	1	1	2	4	6	9	12	16
(*) Denotes	4/0	0	0	1	1	1	3	5	7	10	13
types minus	250	0	0	1	1	1	3	4	6	8	11
outer	300	0	0	1	1	1	2	3	5	7	9
covering	350	0	0	0	1	1	1	3	4	6	8
	400	0	0	0	1	1	1	3	4	6	7
	500	0	0	0	1	1	1	2	3	5	6
	600	0	0	0	0	1	1	1	3	4	5
	700	0	0	0	0	1	1	1	2	3	4
	750	0	0	0	0	1	1	1	2	3	4
	800	0	0	0	0	1	1	1	1	3	4
	1,000	0	0	0	0	0	1	1	1	2	3
THHN,	14	13	22	33	52	76	134	202	291	396	518
THWN,	12	9	16	24	38	56	98	147	212	289	378
THWN-2	10	6	10	15	24	35	62	93	134	182	238
	8	3	6	9	14	20	35	53	77	105	137
	6	2	4	6	10	14	25	38	55	76	99
	4	1	2	4	6	9	16	24	34	46	61
	3	1	1	3	5	7	13	20	29	39	51
	2	1	1	3	4	6	11	17	24	33	43
	1	1	1	1	3	4	8	12	18	24	32

FLEXIBLE METAL CONDUIT—MAXIMUM NUMBER OF CONDUCTORS *(cont.)*

		FMC Trade Size (In.)									
Type	**AWG/kcmil**	**½**	**¾**	**1**	**1¼**	**1½**	**2**	**2½**	**3**	**3½**	**4**
THHN, THWN, THWN-2 *(cont.)*	1/0	1	1	1	2	4	7	10	15	20	27
	2/0	0	1	1	1	3	6	9	12	17	22
	3/0	0	1	1	1	2	5	7	10	14	18
	4/0	0	1	1	1	1	4	6	8	12	15
	250	0	0	1	1	1	3	5	7	9	12
	300	0	0	1	1	1	3	4	6	8	11
	350	0	0	1	1	1	2	3	5	7	9
	400	0	0	0	1	1	1	3	5	6	8
	500	0	0	0	1	1	1	2	4	5	7
	600	0	0	0	0	1	1	1	3	4	5
	700	0	0	0	0	1	1	1	3	4	5
	750	0	0	0	0	1	1	1	2	3	4
	800	0	0	0	0	1	1	1	2	3	4
	1,000	0	0	0	0	0	1	1	1	3	3
FEP, FEPB PFA, PFAH, TFE	14	12	21	32	51	74	130	196	282	385	502
	12	9	15	24	37	54	95	143	206	281	367
	10	6	11	17	26	39	68	103	148	201	263
	8	4	6	10	15	22	39	59	85	115	151
	6	2	4	7	11	16	28	42	60	82	107
	4	1	3	5	7	11	19	29	42	57	75
	3	1	2	4	6	9	16	24	35	48	62
	2	1	1	3	5	7	13	20	29	39	51
PFA, PFAH, TFE	1	1	1	2	3	5	9	14	20	27	36
PFA, PFAH, TFE, Z	1/0	1	1	1	3	4	8	11	17	23	30
	2/0	1	1	1	2	3	6	9	14	19	24
	3/0	0	1	1	1	3	5	8	11	15	20
	4/0	0	1	1	1	2	4	6	9	13	16
Z	14	15	25	39	61	89	157	236	340	463	605
	12	11	18	28	43	63	111	168	241	329	429
	10	6	11	17	26	39	68	103	148	201	263
	8	4	7	11	17	24	43	65	93	127	166
	6	3	5	7	12	17	30	45	65	89	117
	4	1	3	5	8	12	21	31	45	61	80
	3	1	2	4	6	8	15	23	33	45	58
	2	1	1	3	5	7	12	19	27	37	49
	1	1	1	2	4	6	10	15	22	30	39
XHH, XHHW, XHHW-2, ZW	14	9	15	23	36	53	94	141	203	277	361
	12	7	11	18	28	41	72	108	156	212	277
	10	5	8	13	21	30	54	81	116	158	207
	8	3	5	7	11	17	30	45	64	88	115
	6	1	3	5	8	12	22	33	48	65	85
	4	1	2	4	6	9	16	24	34	47	61
	3	1	1	3	5	7	13	20	29	40	52
	2	1	1	3	4	6	11	17	24	33	44
XHH, XHHW, XHHW-2	1	1	1	1	3	5	8	13	18	25	32
	1/0	1	1	1	2	4	7	10	15	21	27
	2/0	0	1	1	2	3	6	9	13	17	23
	3/0	0	1	1	1	3	5	7	10	14	19
	4/0	0	1	1	1	2	4	6	9	12	15
	250	0	0	1	1	1	3	5	7	10	13
	300	0	0	1	1	1	3	4	6	8	11
	350	0	0	1	1	1	2	4	5	7	9
	400	0	0	0	1	1	1	3	5	6	8
	500	0	0	0	1	1	1	3	4	5	7
	600	0	0	0	0	1	1	1	3	4	5
	700	0	0	0	0	1	1	1	3	4	5
	750	0	0	0	0	1	1	1	2	3	4
	800	0	0	0	0	1	1	1	2	3	4

RIGID PVC SCHEDULE 40 CONDUIT—MAXIMUM NUMBER OF CONDUCTORS

		PVC 40 Trade Size (In.)											
Type	**Size**	½	¾	1	1¼	1½	2	2½	3	3½	4	5	6
RHH, RHW, RHW-2	14	4	7	11	20	27	45	64	99	133	171	269	390
	12	3	5	9	16	22	37	53	82	110	142	224	323
	10	2	4	7	13	18	30	43	66	89	115	181	261
	8	1	2	4	7	9	15	22	35	46	60	94	137
	6	1	1	3	5	7	12	18	28	37	48	76	109
	4	1	1	2	4	6	10	14	22	29	37	59	85
	3	1	1	1	4	5	8	12	19	25	33	52	75
	2	1	1	1	3	4	7	10	16	22	28	45	65
	1	0	1	1	1	3	5	7	11	14	19	29	43
	1/0	0	1	1	1	2	4	6	9	13	16	26	37
	2/0	0	0	1	1	1	3	5	8	11	14	22	32
	3/0	0	0	1	1	1	3	4	7	9	12	19	28
	4/0	0	0	1	1	1	2	4	6	8	10	16	24
	250	0	0	0	1	1	1	3	4	6	8	12	18
	300	0	0	0	1	1	1	2	4	5	7	11	16
	350	0	0	0	1	1	1	2	3	5	6	10	14
	400	0	0	0	1	1	1	1	3	4	6	9	13
	500	0	0	0	0	1	1	1	3	4	5	8	11
	600	0	0	0	0	1	1	1	2	3	4	6	9
	700	0	0	0	0	0	1	1	1	3	3	6	8
	750	0	0	0	0	0	1	1	1	2	3	5	8
	800	0	0	0	0	0	1	1	1	2	3	5	7
	1,000	0	0	0	0	0	1	1	1	1	3	4	6
TW	14	8	14	24	42	57	94	135	209	280	361	568	822
	12	6	11	18	32	44	72	103	160	215	277	436	631
	10	4	8	13	24	32	54	77	119	160	206	325	470
	8	2	4	7	13	18	30	43	66	89	115	181	261
RHH*, RHW*, RHW-2*, THHW, THW, THW-2	14	5	9	16	28	38	63	90	139	186	240	378	546
	12	4	8	12	22	30	50	72	112	150	193	304	439
	10	3	6	10	17	24	39	56	87	117	150	237	343
	8	1	3	6	10	14	23	33	52	70	90	142	205
RHH*, RHW*, RHW-2*, TW, THW, THHW, THW-2	6	1	2	4	8	11	18	26	40	53	69	109	157
	4	1	1	3	6	8	13	19	30	40	51	81	117
	3	1	1	3	5	7	11	16	25	34	44	69	100
	2	1	1	2	4	6	10	14	22	29	37	59	85
	1	0	1	1	3	4	7	10	15	20	26	41	60
	1/0	0	1	1	2	3	6	8	13	17	22	35	51
	2/0	0	1	1	1	3	5	7	11	15	19	30	43
Note:	3/0	0	1	1	1	2	4	6	9	12	16	25	36
(*) Denotes	4/0	0	0	1	1	1	3	5	8	10	13	21	30
types minus	250	0	0	1	1	1	3	4	6	8	11	17	25
outer	300	0	0	1	1	1	2	3	5	7	9	15	21
covering	350	0	0	0	1	1	1	3	5	6	8	13	19
	400	0	0	0	1	1	1	3	4	6	7	12	17
	500	0	0	0	1	1	1	2	3	5	6	10	14
	600	0	0	0	0	1	1	1	3	4	5	8	11
	700	0	0	0	0	1	1	1	2	3	4	7	10
	750	0	0	0	0	1	1	1	2	3	4	6	10
	800	0	0	0	0	1	1	1	2	3	4	6	9
	1,000	0	0	0	0	0	1	1	1	2	3	5	7
THHN, THWN, THWN-2	14	11	21	34	60	82	135	193	299	401	517	815	1,178
	12	8	15	25	43	59	99	141	218	293	377	594	859
	10	5	9	15	27	37	62	89	137	184	238	374	541
	8	3	5	9	16	21	36	51	79	106	137	216	312
	6	1	4	6	11	15	26	37	57	77	99	156	225
	4	1	2	4	7	9	16	22	35	47	61	96	138
	3	1	1	3	6	8	13	19	30	40	51	81	117
	2	1	1	3	5	7	11	16	25	33	43	68	98
	1	1	1	1	3	5	8	12	18	25	32	50	73

RIGID PVC SCHEDULE 40 CONDUIT—MAXIMUM NUMBER OF CONDUCTORS *(cont.)*

		PVC 40 Trade Size (In.)											
Type	**Size**	**½**	**¾**	**1**	**1¼**	**1½**	**2**	**2½**	**3**	**3½**	**4**	**5**	**6**
THHN,	1/0	1	1	1	3	4	7	10	15	21	27	42	61
THWN,	2/0	0	1	1	2	3	6	8	13	17	22	35	51
THWN-2	3/0	0	1	1	1	3	5	7	11	14	18	29	42
(cont.)	4/0	0	1	1	1	2	4	6	9	12	15	24	35
	250	0	0	1	1	1	3	4	7	10	12	20	28
	300	0	0	1	1	1	3	4	6	8	11	17	24
	350	0	0	1	1	1	2	3	5	7	9	15	21
	400	0	0	0	1	1	1	3	5	6	8	13	19
	500	0	0	0	1	1	1	2	4	5	7	11	16
	600	0	0	0	1	1	1	1	3	4	5	9	13
	700	0	0	0	0	1	1	1	3	4	5	8	11
	750	0	0	0	0	1	1	1	2	3	4	7	11
	800	0	0	0	0	1	1	1	2	3	4	7	10
	1,000	0	0	0	0	0	1	1	1	3	3	6	8
FEP, FEPB,	14	11	20	33	58	79	131	188	290	389	502	790	1,142
PFA, PFAH,	12	8	15	24	42	58	96	137	212	284	366	577	834
TFE	10	6	10	17	30	41	69	98	152	204	263	414	598
	8	3	6	10	17	24	39	56	87	117	150	237	343
	6	2	4	7	12	17	28	40	62	83	107	169	244
	4	1	3	5	8	12	19	28	43	58	75	118	170
	3	1	2	4	7	10	16	23	36	48	62	98	142
	2	1	1	3	6	8	13	19	30	40	51	81	117
PFA, PFAH, TFE	1	1	1	2	4	5	9	13	20	28	36	56	81
PFA, PFAH,	1/0	1	1	1	3	4	8	11	17	23	30	47	68
TFE, Z	2/0	0	1	1	3	4	6	9	14	19	24	39	56
	3/0	0	1	1	2	3	5	7	12	16	20	32	46
	4/0	0	1	1	1	2	4	6	9	13	16	26	38
Z	14	13	24	40	70	95	158	226	350	469	605	952	1,376
	12	9	17	28	49	68	112	160	248	333	429	675	976
	10	6	10	17	30	41	69	98	152	204	263	414	598
	8	3	6	11	19	26	43	62	96	129	166	261	378
	6	2	4	7	13	18	30	43	67	90	116	184	265
	4	1	3	5	9	12	21	30	46	62	80	126	183
	3	1	2	4	6	9	15	22	34	45	58	92	133
	2	1	1	3	5	7	12	18	28	38	49	77	111
	1	1	1	2	4	6	10	14	23	30	39	62	90
XHH,	14	8	14	24	42	57	94	135	209	280	361	568	822
XHHW,	12	6	11	18	32	44	72	103	160	215	277	436	631
XHHW-2,	10	4	8	13	24	32	54	77	119	160	206	325	470
ZW	8	2	4	7	13	18	30	43	66	89	115	181	261
	6	1	3	5	10	13	22	32	49	66	85	134	193
	4	1	2	4	7	9	16	23	35	48	61	97	140
	3	1	1	3	6	8	13	19	30	40	52	82	118
	2	1	1	3	5	7	11	16	25	34	44	69	99
XHH,	1	1	1	1	3	5	8	12	19	25	32	51	74
XHHW	1/0	1	1	1	3	4	7	10	16	21	27	43	62
XHHW-2	2/0	0	1	1	2	3	6	8	13	17	23	36	52
	3/0	0	1	1	1	3	5	7	11	14	19	30	43
	4/0	0	1	1	1	2	4	6	9	12	15	24	35
	250	0	0	1	1	1	3	5	7	10	13	20	29
	300	0	0	1	1	1	3	4	6	8	11	17	25
	350	0	0	1	1	1	2	3	5	7	9	15	22
	400	0	0	0	1	1	1	3	5	6	8	13	19
	500	0	0	0	1	1	1	2	4	5	7	11	16
	600	0	0	0	1	1	1	1	3	4	5	9	13
	700	0	0	0	0	1	1	1	3	4	5	8	11
	750	0	0	0	0	1	1	1	2	3	4	7	11
	800	0	0	0	0	1	1	1	2	3	4	7	10

RIGID PVC SCHEDULE 80 CONDUIT—MAXIMUM NUMBER OF CONDUCTORS

Type	Size	PVC 80 Trade Size (In.)											
		½	¾	1	1¼	1½	2	2½	3	3½	4	5	6
RHH, RHW, RHW-2	14	3	5	9	17	23	39	56	88	118	153	243	349
	12	2	4	7	14	19	32	46	73	98	127	202	290
	10	1	3	6	11	15	26	37	59	79	103	163	234
	8	1	1	3	6	8	13	19	31	41	54	85	122
	6	1	1	2	4	6	11	16	24	33	43	68	98
	4	1	1	1	3	5	8	12	19	26	33	53	77
	3	0	1	1	3	4	7	11	17	23	29	47	67
	2	0	1	1	3	4	6	9	14	20	25	41	58
	1	0	1	1	1	2	4	6	9	13	17	27	38
	1/0	0	0	1	1	1	3	5	8	11	15	23	33
	2/0	0	0	1	1	1	3	4	7	10	13	20	29
	3/0	0	0	1	1	1	3	4	6	8	11	17	25
	4/0	0	0	0	1	1	2	3	5	7	9	15	21
	250	0	0	0	1	1	1	2	4	5	7	11	16
	300	0	0	0	1	1	1	2	3	5	6	10	14
	350	0	0	0	1	1	1	1	3	4	5	9	13
	400	0	0	0	0	1	1	1	3	4	5	8	12
	500	0	0	0	0	1	1	1	2	3	4	7	10
	600	0	0	0	0	0	1	1	1	3	3	6	8
	700	0	0	0	0	0	1	1	1	2	3	5	7
	750	0	0	0	0	0	1	1	1	2	3	5	7
	800	0	0	0	0	0	1	1	1	2	3	4	7
	1,000	0	0	0	0	0	1	1	1	1	2	4	5
TW	14	6	11	20	35	49	82	118	185	250	324	514	736
	12	5	9	15	27	38	63	91	142	192	248	394	565
	10	3	6	11	20	28	47	67	106	143	185	294	421
	8	1	3	6	11	15	26	37	59	79	103	163	234
RHH*, RHW*, RHW-2*, THHW THW, THW-2	14	4	8	13	23	32	55	79	123	166	215	341	490
	12	3	6	10	19	26	44	63	99	133	173	274	394
	10	2	5	8	15	20	34	49	77	104	135	214	307
	8	1	3	5	9	12	20	29	46	62	81	128	184
RHH*, RHW*, RHW-2*, TW, THW, THHW, THW-2	6	1	1	3	7	9	16	22	35	48	62	98	141
	4	1	1	3	5	7	12	17	26	35	46	73	105
	3	1	1	2	4	6	10	14	22	30	39	63	90
	2	1	1	1	3	5	8	12	19	26	33	53	77
	1	0	1	1	2	3	6	8	13	18	23	37	54
	1/0	0	1	1	1	3	5	7	11	15	20	32	46
	2/0	0	1	1	1	2	4	6	10	13	17	27	39
NOTE: (*) Denotes types minus outer covering	3/0	0	0	1	1	1	3	5	8	11	14	23	33
	4/0	0	0	1	1	1	3	4	7	9	12	19	27
	250	0	0	0	1	1	2	3	5	7	9	15	22
	300	0	0	0	1	1	1	3	5	6	8	13	19
	350	0	0	0	1	1	1	2	4	6	7	12	17
	400	0	0	0	1	1	1	2	4	5	7	10	15
	500	0	0	0	1	1	1	1	3	4	5	9	13
	600	0	0	0	0	1	1	1	2	3	4	7	10
	700	0	0	0	0	1	1	1	2	3	4	6	9
	750	0	0	0	0	0	1	1	1	3	4	6	8
	800	0	0	0	0	0	1	1	1	3	3	6	8
	1,000	0	0	0	0	0	1	1	1	2	3	5	7
THHN, THWN, THWN-2	14	9	17	28	51	70	118	170	265	358	464	736	1,055
	12	6	12	20	37	51	86	124	193	261	338	537	770
	10	4	7	13	23	32	54	78	122	164	213	338	485
	8	2	4	7	13	18	31	45	70	95	123	195	279
	6	1	3	5	9	13	22	32	51	68	89	141	202
	4	1	1	3	6	8	14	20	31	42	54	86	124
	3	1	1	3	5	7	12	17	26	35	46	73	105
	2	1	1	2	4	6	10	14	22	30	39	61	88
	1	0	1	1	3	4	7	10	16	22	29	45	65

RIGID PVC SCHEDULE 80 CONDUIT—MAXIMUM NUMBER OF CONDUCTORS *(cont.)*

		PVC 80 Trade Size (In.)											
Type	**Size**	**1/2**	**3/4**	**1**	**1 1/4**	**1 1/2**	**2**	**2 1/2**	**3**	**3 1/2**	**4**	**5**	**6**
THHN,	1/0	0	1	1	2	3	6	9	14	18	24	38	55
THWN,	2/0	0	1	1	1	3	5	7	11	15	20	32	46
THWN-2	3/0	0	1	1	1	2	4	6	9	13	17	26	38
(cont.)	4/0	0	0	1	1	1	3	5	8	10	14	22	31
	250	0	0	1	1	1	3	4	6	8	11	18	25
	300	0	0	0	1	1	2	3	5	7	9	15	22
	350	0	0	0	1	1	1	3	5	6	8	13	19
	400	0	0	0	1	1	1	3	4	6	7	12	17
	500	0	0	0	1	1	1	2	3	5	6	10	14
	600	0	0	0	0	1	1	1	3	4	5	8	12
	700	0	0	0	0	1	1	1	2	3	4	7	10
	750	0	0	0	0	1	1	1	2	3	4	7	9
	800	0	0	0	0	1	1	1	2	3	4	6	9
	1,000	0	0	0	0	0	1	1	1	2	3	5	7
FEP, FEPB,	14	8	16	27	49	68	115	164	257	347	450	714	1,024
PFA, PFAH,	12	6	12	20	36	50	84	120	188	253	328	521	747
TFE	10	4	8	14	26	36	60	86	135	182	235	374	536
	8	2	5	8	15	20	34	49	77	104	135	214	307
	6	1	3	6	10	14	24	35	55	74	96	152	218
	4	1	2	4	7	10	17	24	38	52	67	106	153
	3	1	1	3	6	8	14	20	32	43	56	89	127
	2	1	1	3	5	7	12	17	26	35	46	73	105
PFA, PFAH, TFE	1	1	1	1	3	5	8	11	18	25	32	51	73
PFA, PFAH,	1/0	0	1	1	3	4	7	10	15	20	27	42	61
TFE, Z	2/0	0	1	1	2	3	5	8	12	17	22	35	50
	3/0	0	1	1	1	2	4	6	10	14	18	29	41
	4/0	0	0	1	1	1	4	5	8	11	15	24	34
Z	14	10	19	33	59	82	138	198	310	418	542	860	1,233
	12	7	14	23	42	58	98	141	220	297	385	610	875
	10	4	8	14	26	36	60	86	135	182	235	374	536
	8	3	5	9	16	22	38	54	85	115	149	236	339
	6	2	4	6	11	16	26	38	60	81	104	166	238
	4	1	2	4	8	11	18	26	41	55	72	114	164
	3	1	2	3	5	8	13	19	30	40	52	83	119
	2	1	1	2	5	6	11	16	25	33	43	69	99
	1	0	1	2	4	5	9	13	20	27	35	56	80
XHH,	14	6	11	20	35	49	82	118	185	250	324	514	736
XHHW,	12	5	9	15	27	38	63	91	142	192	248	394	565
XHHW-2,	10	3	6	11	20	28	47	67	106	143	185	294	421
ZW	8	1	3	6	11	15	26	37	59	79	103	163	234
	6	1	2	4	8	11	19	28	43	59	76	121	173
	4	1	1	3	6	8	14	20	31	42	55	87	125
	3	1	1	3	5	7	12	17	26	36	47	74	106
	2	1	1	2	4	6	10	14	22	30	39	62	89
XHH, XHHW,	1	0	1	1	3	4	7	10	16	22	29	46	66
XHHW-2	1/0	0	1	1	2	3	6	9	14	19	24	39	56
	2/0	0	1	1	1	3	5	7	11	16	20	32	46
	3/0	0	1	1	1	2	4	6	9	13	17	27	38
	4/0	0	0	1	1	1	3	5	8	11	14	22	32
	250	0	0	1	1	1	3	4	6	9	11	18	26
	300	0	0	1	1	1	2	3	5	7	10	15	22
	350	0	0	0	1	1	1	3	5	6	8	14	20
	400	0	0	0	1	1	1	3	4	6	7	12	17
	500	0	0	0	1	1	1	2	3	5	6	10	14
	600	0	0	0	0	1	1	1	3	4	5	8	11
	700	0	0	0	0	1	1	1	2	3	4	7	10
	750	0	0	0	0	1	1	1	2	3	4	6	9
	800	0	0	0	0	1	1	1	1	3	4	6	9

COPPER BUS-BAR DATA
Sizes, Weights and Resistances

Thickness (Inches)	Width (Inches)	Wts. per Ft. at .3213 Lbs. per Cubic In.	Area in Square In.	Ohms per Ft. at 8.341 per Sq. Mil. Ft.	Capacity in Amperes
1/16	1/2	.1205	.0313	.00026691	30
1/16	3/4	.1807	.0469	.00017790	50
1/16	1	.2410	.0625	.00013344	60
1/16	1-1/2	.3615	.0938	.00008897	90
1/8	1/2	.2410	.0625	.00013344	75
1/8	3/4	.3615	.0938	.00008897	90
1/8	1	.4820	.125	.00006672	125
1/8	1-1/2	.7230	.1875	.00004448	200
1/8	2	.9640	.25	.00003336	250
1/4	3/4	.7230	.1875	.00004448	185
1/4	1	.9640	.25	.00003336	250
1/4	1-1/4	1.205	.3125	.00002669	315
1/4	1-1/2	1.446	.375	.00002224	375
1/4	1-3/4	1.687	.4375	.00001906	435
1/4	2	1.928	.5	.00001668	500
1/4	2-1/4	2.169	.5625	.00001482	565
1/4	2-1/2	2.410	.625	.00001334	630
1/2	3/4	1.446	.375	.00002224	370
1/2	1	1.928	.500	.00001668	500
1/2	1-1/4	2.410	.625	.00001334	625
1/2	1-1/2	2.892	.750	.00001112	750
1/2	1-3/4	3.374	.875	.00000953	875
1/2	2	3.856	1.	.00000834	1000
1/2	2-1/4	4.338	1.125	.00000741	1185
1/2	2-1/2	4.820	1.25	.00000667	1250
1/2	2-3/4	5.304	1.375	.00000606	1375
1/2	3	5.784	1.500	.00000556	1500
1/2	3-1/4	6.266	1.625	.00000513	1625
1/2	3-1/2	6.748	1.750	.00000475	1750
1/2	3-3/4	7.23	1.875	.00000444	1875
1/2	4	7.712	2.000	.00000417	2000
3/4	1	2.892	.750	.00001112	750
3/4	1-1/2	4.338	1.125	.00000741	1125
3/4	2	5.784	1.500	.00000556	1500
3/4	2-1/2	7.23	1.875	.00000444	1875
3/4	3	8.676	2.250	.00000370	2250
3/4	3-1/2	10.122	2.625	.00000317	2650
3/4	4	11.568	3.000	.00000278	3000

Carrying capacity is figured at 1,000 amperes per square inch.

COMPARATIVE WEIGHTS OF COPPER AND ALUMINUM CONDUCTORS/Lbs. per 1,000 Ft.						
Wire Size	Bare Solid			Bare Stranded		
	Copper	Alum.	Diff.	Copper	Alum.	Diff.
18	4.92	1.49	3.43	5.02	1.53	3.49
16	7.82	2.38	5.44	7.97	2.43	5.54
14	12.43	3.78	8.65	12.68	3.86	8.82
12	19.77	6.01	13.76	20.16	6.13	14.03
10	31.43	9.55	21.88	32.06	9.75	22.31
8	50.0	15.2	34.8	51.0	15.5	35.5
6	79.5	24.2	55.3	81.0	24.6	56.4
4	126.4	38.4	88.0	128.9	39.2	89.7
2	200.9	60.8	140.1	204.9	62.3	142.6
1	253.3	77.0	176.3	258.4	78.6	179.8
1/0	319.5	97.2	222.3	325.8	99.1	226.7
2/0	402.8	122.6	280.2	410.9	124.9	286.0
3/0	507.9	154.6	353.3	518.1	157.5	360.6
4/0	640.5	194.9	445.6	653.3	198.6	454.7
250	—	—	—	771.9	234.7	537.2
300	—	—	—	926.3	281.6	644.7
350	—	—	—	1081	328.6	752
400	—	—	—	1235	375.5	859
450	—	—	—	1389	422.4	967
500	—	—	—	1544	469.4	1075
600	—	—	—	1853	563	1290
700	—	—	—	2161	657	1504
750	—	—	—	2316	704	1612
800	—	—	—	2470	751	1719
900	—	—	—	2779	845	1934
1000	—	—	—	3088	939	2149
1250	—	—	—	3859	1173	2686
1500	—	—	—	4631	1410	3221
1750	—	—	—	5403	1645	3758
2000	—	—	—	6175	1880	4295

COPPER WIRE SPECIFICATIONS				
Wire Size (AWG)	Area (Circular Mils)	Diameter (Mils, 1000th In.)	Diameter (mm)	Weight (lbs. per 1000 ft.)
40	9.9	3.1	.080	.0200
38	15.7	4	.101	.0476
36	25	5	.127	.0757
34	39.8	6.3	.160	.120
32	63.2	8	.202	.191
30	101	10	.255	.304
28	160	12.6	.321	.484
26	254	15.9	.405	.769
24	404	20.1	.511	1.22
22	642	25.3	.644	1.94
20	1,020	32	.812	3.09
18	1,620	40	1.024	4.92
16	2,580	51	1.291	7.82
14	4,110	64	1.628	12.4
12	6,530	81	2.053	19.8
10	10,400	102	2.588	31.4
8	16,500	128	3.264	50
6	26,300	162	4.115	79.5
4	41,700	204	5.189	126
3	52,600	229	5.827	159
2	66,400	258	6.544	201
1	83,700	289	7.348	253
1/0	106,000	325	8.255	319
2/0	133,000	365	9.271	403
3/0	168,000	410	10.414	508
4/0	212,000	460	11.684	641

COPPER WIRE RESISTANCE

Wire Size (AWG)	77°F (ft. per ohm)	149°F (ft. per ohm)	77°F (ohms per 1,000 ft.)	149°F (ohms per 1,000 ft.)
40	.93	.81	1,070	1,230
38	1.5	1.3	673	776
36	2.4	2.0	423	488
34	3.8	3.3	266	307
32	6.0	5.2	167	193
30	9.5	8.3	105	121
28	15.1	13.1	66.2	76.4
26	24.0	20.8	41.6	48.0
24	38.2	33.1	26.2	30.2
22	60.6	52.6	16.5	19.0
20	96.2	84.0	10.4	11.9
18	153.6	133.2	6.51	7.51
16	244.5	211.4	4.09	4.73
14	387.6	336.7	2.58	2.97
12	617.3	534.8	1.62	1.87
10	980.4	847.5	1.02	1.18
8	1,560	1,353	.641	.739
6	2,481	2,151	.403	.465
4	3,953	3,425	.253	.292
3	4,975	4,310	.201	.232
2	6,289	5,435	.159	.184
1	7,936	6,849	.126	.146
1/0	10,000	8,621	.100	.116
2/0	12,658	10,870	.079	.092
3/0	15,873	13,699	.063	.073
4/0	20,000	17,544	.050	.057

CONDUCTOR DIMENSIONS AND RESISTANCES								
Wire Size	Area (Cir. Mils)	Stranding		Overall Dimensions		DC Resistance Ohm/K Ft. @ 75°C		
						Copper		Alum.
		Wire Qty	Dia. (In.)	Dia. (In.)	Area ($In.^2$)	Uncoated	Coated	
18	1620	1	—	0.040	0.001	7.77	8.08	12.8
18	1620	7	0.015	0.046	0.002	7.95	8.45	13.1
16	2580	1	—	0.051	0.002	4.89	5.08	8.05
16	2580	7	0.019	0.058	0.003	4.99	5.29	8.21
14	4110	1	—	0.064	0.003	3.07	3.19	5.06
14	4110	7	0.024	0.073	0.004	3.14	3.26	5.17
12	6530	1	—	0.081	0.005	1.93	2.01	3.18
12	6530	7	0.030	0.092	0.006	1.98	2.05	3.25
10	10380	1	—	0.102	0.008	1.21	1.26	2.00
10	10380	7	0.038	0.116	0.011	1.24	1.29	2.04
8	16510	1	—	0.128	0.013	0.764	0.786	1.26
8	16510	7	0.049	0.146	0.017	0.778	0.809	1.28

CONDUCTOR DIMENSIONS AND RESISTANCES *(cont.)*

Wire Size	Area (Cir. Mils)	Stranding		Overall Dimensions		DC Resistance Ohm/K Ft. @ 75°C		
						Copper		Alum.
		Wire Qty	Dia. (In.)	Dia. (In.)	Area (In.^{2})	Uncoated	Coated	
6	26240	7	0.061	0.184	0.027	0.491	0.510	0.808
4	41740	7	0.077	0.232	0.042	0.308	0.321	0.508
3	52620	7	0.087	0.260	0.053	0.245	0.254	0.403
2	66360	7	0.097	0.292	0.067	0.194	0.201	0.319
1	83690	19	0.066	0.332	0.087	0.154	0.160	0.253
1/0	105600	19	0.074	0.372	0.109	0.122	0.127	0.201
2/0	133100	19	0.084	0.418	0.137	0.0967	0.101	0.159
3/0	167800	19	0.094	0.470	0.173	0.0766	0.0797	0.126
4/0	211600	19	0.106	0.528	0.219	0.0608	0.0626	0.100
250	250000	37	0.082	0.575	0.260	0.0515	0.0535	0.0847
300	300000	37	0.090	0.630	0.312	0.0429	0.0446	0.0707
350	350000	37	0.097	0.681	0.364	0.0367	0.0382	0.0605

CONDUCTOR DIMENSIONS AND RESISTANCES *(cont.)*								
Wire Size	Area (Cir. Mils)	Stranding		Overall Dimensions		DC Resistance Ohm/K Ft. @ 75°C		
						Copper		Alum.
		Wire Qty	Dia. (In.)	Dia. (In.)	Area (In.^{2})	Uncoated	Coated	
400	400000	37	0.104	0.728	0.416	0.0321	0.0331	0.0529
500	500000	37	0.116	0.813	0.519	0.0258	0.0265	0.0424
600	600000	61	0.099	0.893	0.626	0.0214	0.0223	0.0353
700	700000	61	0.107	0.964	0.730	0.0184	0.0189	0.0303
750	750000	61	0.111	0.998	0.782	0.0171	0.0176	0.0282
800	800000	61	0.114	1.030	0.834	0.0161	0.0166	0.0265
900	900000	61	0.122	1.094	0.940	0.0143	0.0147	0.0235
1000	1000000	61	0.128	1.152	1.042	0.0129	0.0132	0.0212
1250	1250000	91	0.117	1.289	1.305	0.0103	0.0106	0.0169
1500	1500000	91	0.128	1.412	1.566	0.00858	0.00883	0.0141
1750	1750000	127	0.117	1.526	1.829	0.00735	0.00756	0.0121
2000	2000000	127	0.126	1.632	2.092	0.00643	0.00662	0.0106

AC RESISTANCE, REACTANCE, & IMPEDANCE FOR 600 VOLT CABLES MEASURED IN OHMS TO NEUTRAL PER 1000 FEET, 3 PHASE, 60 Hz, 75°C BASED ON THREE SINGLE CONDUCTORS IN PVC OR STEEL CONDUIT

Wire Size	AC Resistance for Uncoated Copper Wires		AC Resistance for Aluminum Wires		Reactance for All Wires		Impedance at 0.85 PF for Uncoated Copper Wires		Impedance at 0.85 PF for Aluminum Wires	
	PVC	Steel	PVC	Steel	PVC	Steel	PVC	Steel	PVC	Steel
14	3.1	3.1	—	—	0.058	0.073	2.7	2.7	—	—
12	2.0	2.0	3.2	3.2	0.054	0.068	1.7	1.7	2.8	2.8
10	1.2	1.2	2.0	2.0	0.050	0.063	1.1	1.1	1.8	1.8
8	0.78	0.78	1.3	1.3	0.052	0.065	0.69	0.70	1.1	1.1
6	0.49	0.49	0.81	0.81	0.051	0.064	0.44	0.45	0.71	0.72
4	0.31	0.31	0.51	0.51	0.048	0.060	0.29	0.30	0.46	0.46
3	0.25	0.25	0.40	0.40	0.047	0.059	0.23	0.24	0.37	0.37

AC RESISTANCE, REACTANCE, & IMPEDANCE FOR 600 VOLT CABLES MEASURED IN OHMS TO NEUTRAL PER 1000 FEET, 3 PHASE, 60 Hz, 75°C BASED ON THREE SINGLE CONDUCTORS IN PVC OR STEEL CONDUIT *(cont.)*

Wire Size	AC Resistance for Uncoated Copper Wires		AC Resistance for Aluminum Wires		Reactance for All Wires		Impedance at 0.85 PF for Uncoated Copper Wires		Impedance at 0.85 PF for Aluminum Wires	
	PVC	Steel	PVC	Steel	PVC	Steel	PVC	Steel	PVC	Steel
2	0.19	0.20	0.32	0.32	0.045	0.057	0.19	0.20	0.30	0.30
1	0.15	0.16	0.25	0.25	0.046	0.057	0.16	0.16	0.24	0.25
1/0	0.12	0.12	0.20	0.20	0.044	0.055	0.13	0.13	0.19	0.20
2/0	0.10	0.10	0.16	0.16	0.043	0.054	0.11	0.11	0.16	0.16
3/0	0.077	0.079	0.13	0.13	0.042	0.052	0.088	0.094	0.13	0.14
4/0	0.062	0.063	0.10	0.10	0.041	0.051	0.074	0.080	0.11	0.11
250	0.052	0.054	0.085	0.086	0.041	0.052	0.066	0.073	0.094	0.10

AC RESISTANCE, REACTANCE, & IMPEDANCE FOR 600 VOLT CABLES MEASURED IN OHMS TO NEUTRAL PER 1000 FEET, 3 PHASE, 60 Hz, 75°C BASED ON THREE SINGLE CONDUCTORS IN PVC OR STEEL CONDUIT *(cont.)*

Wire Size	AC Resistance for Uncoated Copper Wires		AC Resistance for Aluminum Wires		Reactance for All Wires		Impedance at 0.85 PF for Uncoated Copper Wires		Impedance at 0.85 PF for Aluminum Wires	
	PVC	Steel	PVC	Steel	PVC	Steel	PVC	Steel	PVC	Steel
300	0.044	0.045	0.071	0.072	0.041	0.051	0.059	0.065	0.082	0.088
350	0.038	0.039	0.061	0.063	0.040	0.050	0.053	0.060	0.073	0.080
400	0.033	0.035	0.054	0.055	0.040	0.049	0.049	0.056	0.066	0.073
500	0.027	0.029	0.043	0.045	0.039	0.048	0.043	0.050	0.057	0.064
600	0.023	0.025	0.036	0.038	0.039	0.048	0.040	0.047	0.051	0.058
750	0.019	0.021	0.029	0.031	0.038	0.048	0.036	0.043	0.045	0.052
1000	0.015	0.018	0.023	0.025	0.037	0.046	0.032	0.040	0.039	0.046

AMPACITIES OF COPPER CONDUCTORS (1)

Ampacities of Single Insulated Conductors Rated 0–2000 Volts in Free Air

Ambient Temperature 30°C (86°F)				Ambient Temperature 40°C (104°F)			
Wire Size	60°C (140°F)	75°C (167°F)	90°C (194°F)	150°C (302°F)	200°C (392°F)	250°C (482°F)	Bare Conductors with Max. Temp. 80°C (176°F)
AWG kcmil	Types	Types	Types	Type	Types	Types	
	TW UF	RHW, THHW, THW, THWN, XHHW, ZW	TBS, SA, SIS, FEP, MI, FEPB, ZW-2, RHH, RHW-2, THHN, THHW, THW-2, XHH, USE-2, THWN-2 XHHW, XHHW-2.	Z	FEP, FEPB, PFA SA	PFAH, TFE Nickel or nickel-coated copper	
14*	25	30	35	46	54	59	—
12*	30	35	40	60	68	78	—
10*	40	50	55	80	90	107	—
8	60	70	80	106	124	142	98
6	80	95	105	155	165	205	124
4	105	125	140	190	220	278	155
3	120	145	165	214	252	327	—
2	140	170	190	255	293	381	209
1	165	195	220	293	344	440	—
1/0	195	230	260	339	399	532	282
2/0	225	265	300	390	467	591	329
3/0	260	310	350	451	546	708	382
4/0	300	360	405	529	629	830	444
250	340	405	455	—	—	—	494

Temperature Correction Factors

Ambient Temp.°C	For Other Than 30°C			For Other Than 40°C			Ambient Temp.°C
	Multiply the Ampacities above by the Factors Below						
21–25	1.08	1.05	1.04	.95	.97	.98	41–50
26–30	1.00	1.00	1.00	.90	.94	.95	51–60
31–35	.91	.94	.96	.85	.90	.93	61–70
36–40	.82	.88	.91	.80	.87	.90	71–80
41–45	.71	.82	.87	.74	.83	.87	81–90
46–50	.58	.75	.82	.67	.79	.85	91–100
51–55	.41	.67	.76	.52	.71	.79	101–120
56–60	—	.58	.71	.30	.61	.72	121–140
61–70	—	.33	.58	—	.50	.65	141–160
71–80	—	—	.41	—	.35	.58	161–180
				—	—	.49	181–200
				—	—	.35	201–225

*Unless specifically permitted by the NEC®, overcurrent protection for copper conductors shall not exceed 15 amps for no. 14 AWG, 20 amps for no. 12 AWG, and 30 amps for no. 10 AWG.

AMPACITIES OF COPPER CONDUCTORS (1) *(cont.)*

Ampacities of Single Insulated Conductors Rated 0–2000 Volts in Free Air

	Ambient Temperature 30°C (86°F)			Ambient Temperature 40°C (104°F)			
Wire Size	**60°C (140°F)**	**75°C (167°F)**	**90°C (194°F)**	**150°C (302°F)**	**200°C (392°F)**	**250°C (482°F)**	**Bare Conductors with Max. Temp. 80°C (176°F)**
	Types	**Types**	**Types**	**Type**	**Types**	**Types**	
kcmil	TW, UF	RHW, THHW, THW, THWN, XHHW, ZW	TBS, SA, SIS, FEP, MI, FEPB, ZW-2, RHH, RHW-2, THHN, THHW, THW-2, XHH, USE-2, THWN-2 XHHW, XHHW-2.	Z	FEP, FEPB, PFA SA	PFAH, TFE Nickel or nickel-coated copper	
300	375	445	505	—	—	—	556
350	420	505	570	—	—	—	—
400	455	545	615	—	—	—	—
500	515	620	700	—	—	—	773
600	575	690	780	—	—	—	
700	630	755	855	—	—	—	—
750	655	785	885	—	—	—	1,000
800	680	815	920	—	—	—	—
900	730	870	985	—	—	—	—
1,000	780	935	1,055	—	—	—	1,193
1,250	890	1,065	1,200	—	—	—	—
1,500	980	1,175	1,325	—	—	—	—
1,750	1,070	1,280	1,445	—	—	—	—
2,000	1,155	1,385	1,560	—	—	—	—

Temperature Correction Factors

Ambient Temp.°C	For Other Than 30°C Multiply the Ampacities above by the Factors Below		
21–25	1.08	1.05	1.04
26–30	1.00	1.00	1.00
31–35	.91	.94	.96
36–40	.82	.88	.91
41–45	.71	.82	.87
46–50	.58	.75	.82
51–55	.41	.67	.76
56–60	—	.58	.71
61–70	—	.33	.58
71–80	—	—	.41

AMPACITIES OF COPPER CONDUCTORS (3)

Ampacities of Not More Than 3 Insulated Conductors Rated 0–2000 Volts							
In Cable, Raceway, or Earth, Ambient Temperature 30°C (86°F)				In Cable or Raceway, Ambient Temperature 40°C (104°F)			
Wire Size	60°C (140°F)	75°C (167°F)	90°C (194°F)	150°C (302°F)	200°C (392°F)	250°C (482°F)	Wire Size
AWG kcmil	Types TW UF	Types RHW, THHW, THW, THWN, XHHW, USE, ZW	Types TBS, SA, SIS, FEP, MI FEPB, ZW-2, RHH, RHW-2, THHN, THHW, THW-2, XHH, USE-2, THWN-2 XHHW, XHHW-2.	Type Z	Types FEP, FEPB, PFA SA	Types PFAH, TFE / Nickel or nickel-coated copper	AWG
14*	20	20	25	34	36	39	14
12*	25	25	30	43	45	54	12
10*	30	35	40	55	60	73	10
8	40	50	55	76	83	93	8
6	55	65	75	96	110	117	6
4	70	85	95	120	125	148	4
3	85	100	110	143	152	166	3
2	95	115	130	160	171	191	2
1	110	130	150	186	197	215	1
1/0	125	150	170	215	229	244	1/0
2/0	145	175	195	251	260	273	2/0
3/0	165	200	225	288	297	308	3/0
4/0	195	230	260	332	346	361	4/0
250	215	255	290	—	—	—	—

Temperature Correction Factors							
Ambient Temp.°C	For Other Than 30°C			For Other Than 40°C			Ambient Temp.°C
	Multiply the Ampacities above by the Factors Below						
21–25	1.08	1.05	1.04	.95	.97	.98	41–50
26–30	1.00	1.00	1.00	.90	.94	.95	51–60
31–35	.91	.94	.96	.85	.90	.93	61–70
36–40	.82	.88	.91	.80	.87	.90	71–80
41–45	.71	.82	.87	.74	.83	.87	81–90
46–50	.58	.75	.82	.67	.79	.85	91–100
51–55	.41	.67	.76	.52	.71	.79	101–120
56–60	—	.58	.71	.30	.61	.72	121–140
61–70	—	.33	.58	—	.50	.65	141–160
71–80	—	—	.41	—	.35	.58	161–180
				—	—	.49	181–200
				—	—	.35	201–225

*Unless specifically permitted by the NEC®, overcurrent protection for copper conductors shall not exceed 15 amps for no. 14 AWG, 20 amps for no. 12 AWG, and 30 amps for no. 10 AWG.

AMPACITIES OF COPPER CONDUCTORS (3) *(cont.)*

Ampacities of Not More Than 3 Insulated Conductors Rated 0–2000 Volts							
In Cable, Raceway, or Earth, Ambient Temperature 30°C (86°F)				In Cable or Raceway, Ambient Temperature 40°C (104°F)			
Wire Size	60°C (140°F)	75°C (167°F)	90°C (194°F)	150°C (302°F)	200°C (392°F)	250°C (482°F)	Wire Size
kcmil	Types TW, UF	Types RHW, THHW, THW, THWN, XHHW, USE, ZW	Types TBS, SA, SIS, FEP, MI FEPB, ZW-2, RHH, RHW-2, THHN, THHW, THW-2, XHH, USE-2, THWN-2 XHHW, XHHW-2.	Type Z	Types FEP, FEPB, PFA SA	Types PFAH, TFE Nickel or nickel-coated copper	
300	240	285	320				
350	260	310	350				
400	280	335	380				
500	320	380	430				
600	355	420	475				
700	385	460	520				
750	400	475	535				
800	410	490	555				
900	435	520	585				
1,000	455	545	615				
1,250	495	590	665				
1,500	520	625	705				
1,750	545	650	735				
2,000	560	665	750				
Temperature Correction Factors							
Ambient Temp.°C	For Other Than 30°C Multiply the Ampacities above by the Factors Below						
21–25	1.08	1.05	1.04				
26–30	1.00	1.00	1.00				
31–35	.91	.94	.96				
36–40	.82	.88	.91				
41–45	.71	.82	.87				
46–50	.58	.75	.82				
51–55	.41	.67	.76				
56–60	—	.58	.71				
61–70	—	.33	.58				
71–80	—	—	.41				

AMPACITIES OF ALUMINUM OR COPPER-CLAD ALUMINUM CONDUCTORS (1)

Ampacities of Single Insulated Conductors Rated 0–2000 Volts in Free Air

Ambient Temperature 30°C (86°F)				Ambient Temp. 40°C (104°F)	
Wire Size	**60°C (140°F)**	**75°C (167°F)**	**90°C (194°F)**	**150°C (302°F)**	**Wire Size**
AWG kcmil	**Types** TW, UF	**Types** THWN, THHW, XHHW, RHW, THW	**Types** TBS, XHH, RHH, RHW-2, XHHW, USE-2, THHN, SA, THHW, THW-2, XHHW-2, ZW-2, THWN-2, SIS	**Type** Z	**AWG**
12*	25	30	35	47	12
10*	35	40	40	63	10
8	45	55	60	83	8
6	60	75	80	112	6
4	80	100	110	148	4
3	95	115	130	170	3
2	110	135	150	198	2
1	130	155	175	228	1
1/0	150	180	205	263	1/0
2/0	175	210	235	305	2/0
3/0	200	240	275	351	3/0
4/0	235	280	315	411	4/0
250	265	315	355		

Temperature Correction Factors

Ambient Temp.°C	For Other Than 30°C			40°C	Ambient Temp.°C
	Multiply the Ampacities above by the Factors Below				
21–25	1.08	1.05	1.04	.95	41–50
26–30	1.00	1.00	1.00	.90	51–60
31–35	.91	.94	.96	.85	61–70
36–40	.82	.88	.91	.80	71–80
41–45	.71	.82	.87	.74	81–90
46–50	.58	.75	.82	.67	91–100
51–55	.41	.67	.76	.52	101–120
56–60	—	.58	.71	.30	121–140
61–70	—	.33	.58		
71–80	—	—	.41		

*Unless specifically permitted by the NEC®, overcurrent protection for aluminum and copper-clad aluminum conductors shall not exceed 15 amps for no. 12 AWG and 25 amps for no. 10 AWG.

AMPACITIES OF ALUMINUM OR COPPER-CLAD ALUMINUM CONDUCTORS (1) *(cont.)*

Ampacities of Single Insulated Conductors Rated 0–2000 Volts in Free Air					
	Ambient Temperature 30°C (86°F)			Ambient Temp. 40°C (104°F)	
Wire Size	**60°C (140°F)**	**75°C (167°F)**	**90°C (194°F)**	**150°C (302°F)**	**Wire Size**
	Types	**Types**	**Types**	**Type**	
kcmil	TW, UF	THWN, THHW, XHHW, RHW, THW	TBS, XHH RHH, RHW-2, XHHW, USE-2, THHN, SA, THHW, THW-2, XHHW-2, ZW-2, THWN-2, SIS	Z	
300	290	350	395		
350	330	395	445		
400	355	425	480		
500	405	485	545		
600	455	540	615		
700	500	595	675		
750	515	620	700		
800	535	645	725		
900	580	700	785		
1,000	625	750	845		
1,250	710	855	960		
1,500	795	950	1,075		
1,750	875	1,050	1,185		
2,000	960	1,150	1,335		
	Temperature Correction Factors				
Ambient Temp.°C	**For Other Than 30°C Multiply the Ampacities above by the Factors Below**				
21–25	1.08	1.05	1.04		
26–30	1.00	1.00	1.00		
31–35	.91	.94	.96		
36–40	.82	.88	.91		
41–45	.71	.82	.87		
46–50	.58	.75	.82		
51–55	.41	.67	.76		
56–60	—	.58	.71		
61–70	—	.33	.58		
71–80	—	—	.41		

AMPACITIES OF ALUMINUM AND COPPER-CLAD ALUMINUM CONDUCTORS (3)

Ampacities of Not More Than 3 Insulated Conductors Rated 0–2000 Volts					
	In Cable, Raceway, or Earth, Ambient Temperature 30°C (86°F)			In Cable or Raceway, Ambient Temp. 40°C (104°F)	
Wire Size	60°C (140°F)	75°C (167°F)	90°C (194°F)	150°C (302°F)	Wire Size
AWG kcmil	Types	Types	Types	Type	AWG
	TW, UF	THWN, THHW, XHHW, USE, THW, RHW	TBS, SA, SIS, THHN, THHW, THW-2, THWN-2, RHH, RHW-2, USE-2, XHH, XHHW, XHHW-2, ZW-2	Z	
12*	20	20	25	30	12
10*	25	30	35	44	10
8	30	40	45	57	8
6	40	50	60	75	6
4	55	65	75	94	4
3	65	75	85	109	3
2	75	90	100	124	2
1	85	100	115	145	1
1/0	100	120	135	169	1/0
2/0	115	135	150	198	2/0
3/0	130	155	175	227	3/0
4/0	150	180	205	260	4/0
250	170	205	230		
Temperature Correction Factors					
Ambient Temp.°C	For Other Than 30°C			40°C	Ambient Temp.°C
	Multiply the Ampacities above by the Factors Below				
21–25	1.08	1.05	1.04	.95	41–50
26–30	1.00	1.00	1.00	.90	51–60
31–35	.91	.94	.96	.85	61–70
36–40	.82	.88	.91	.80	71–80
41–45	.71	.82	.87	.74	81–90
46–50	.58	.75	.82	.67	91–100
51–55	.41	.67	.76	.52	101–120
56–60	—	.58	.71	.30	121–140
61–70	—	.33	.58		
71–80	—	—	.41		

*Unless specifically permitted by the NEC®, overcurrent protection for aluminum and copper-clad aluminum conductors shall not exceed 15 amps for no. 12 AWG and 25 amps for no. 10 AWG.

AMPACITIES OF ALUMINUM AND COPPER-CLAD ALUMINUM CONDUCTORS (3) *(cont.)*

Ampacities of Not More Than 3 Insulated Conductors Rated 0–2000 Volts					
	In Cable, Raceway, or Earth, Ambient Temperature 30°C (86°F)			In Cable or Raceway Ambient Temp. 40°C (104°F)	
Wire Size	60°C (140°F)	75°C (167°F)	90°C (194°F)	150°C (302°F)	Wire Size
kcmil	Types	Types	Types	Type	
	TW, UF	THWN, THHW, XHHW, USE, THW, RHW	TBS, SA, SIS, THHN, THHW, THW-2, THWN-2, RHH, RHW-2, USE-2, XHH, XHHW, XHHW-2, ZW-2	Z	
300	190	230	255		
350	210	250	280		
400	225	270	305		
500	260	310	350		
600	285	340	385		
700	310	375	420		
750	320	385	435		
800	330	395	450		
900	355	425	480		
1,000	375	445	500		
1,250	405	485	545		
1,500	435	520	585		
1,750	455	545	615		
2,000	470	560	630		
Temperature Correction Factors					
Ambient Temp.°C	For Other Than 30°C Multiply the Ampacities above by the Factors Below				
21–25	1.08	1.05	1.04		
26–30	1.00	1.00	1.00		
31–35	.91	.94	.96		
36–40	.82	.88	.91		
41–45	.71	.82	.87		
46–50	.58	.75	.82		
51–55	.41	.67	.76		
56–60	—	.58	.71		
61–70	—	.33	.58		
71–80	—	—	.41		

AMPACITY RATINGS FOR FLEXIBLE CORDS AND POWER CABLES AT 30°C AMBIENT TEMPERATURE			
Wire Size (AWG)	Types C, E, S, SJ, SJO, SJOW, SJOO, SO, SOW, SOO, SP-1, SP-2, SP-3, SRD, SJT, SJTW, SPE-1, SPE-2, SPE-3, SPT-1, SPT-1W, SPT-2, SPT-2W, SPT-3, ST, SRDE, SRDT, STO, STOW, STOO, STOOW, SVE, SVEO, SVT, SVTO, SVTOO		Types HPD, HPN, HSJ, HSJO, HSJOC
	A*	B*	
18	7	10	10
17	—	12	13
16	10	13	15
15	—	—	17
14	15	18	20
12	20	25	30
10	25	30	35
8	35	40	—
6	45	55	—
4	60	70	—
2	80	95	—

A* For Three (3) current-carrying conductors.

B* For Two (2) current-carrying conductors.

AMPACITY ADJUSTMENTS FOR 4 OR MORE CONDUCTORS IN A CABLE OR RACEWAY

Number of Current-Carrying Conductors	Percent of Values in Ampacity Charts/Adjust for Ambient Temperature (if Necessary)
4 to 6	80%
7 to 9	70%
10 to 20	50%
21 to 30	45%
31 to 40	40%
41 and more	35%

Note: For use with ampacity charts on pages 3–42 through 3–50.

AMPACITY RATINGS FOR SINGLE-PHASE SERVICE OR FEEDER CONDUCTORS IN NORMAL DWELLING UNITS

Service or Feeder Rating in Amps	Copper	Aluminum or Copper-Clad Aluminum
100	4 AWG	2 AWG
110	3	1
125	2	1/0
150	1	2/0
175	1/0	3/0
200	2/0	4/0
225	3/0	250 kcmil
250	4/0	300
300	250 kcmil	350
350	350	500
400	400	600

BRANCH CIRCUIT REQUIREMENTS					
Branch Circuit Rating in Amps	**Minimum Circuit Conductor Size**	**Minimum Tap Conductor Size**	**Overcurrent Protection Rating in Amps**	**Basic Electrical Lampholder Type**	**Outlet Receptacle Rating in Amps**
15	14 AWG	14 AWG	15	Any	15 max
20	12 AWG	14 AWG	20	Any	15 or 20
30	10 AWG	14 AWG	30	Heavy Duty	30
40	8 AWG	12 AWG	40	Heavy Duty	40 or 50
50	6 AWG	12 AWG	50	Heavy Duty	50

BRANCH CIRCUIT PERMISSIBLE LOADS

The load shall never exceed the branch-circuit amp rating. An individual branch circuit shall be permitted to supply any load for which it is rated. A branch circuit supplying two or more outlets or receptacles shall supply only the loads specified according to its size as follows:

(1) 15- or 20-amp branch circuits shall be permitted to supply lighting, other equipment, or a combination of both, except small-appliance branch circuits, laundry branch circuits, and bathroom branch circuits required in dwelling units and shall supply only those receptacle outlets.

   (A) The rating of any one cord-and-plug-connected utilization equipment not fastened in place shall not exceed 80 percent of the branch-circuit amp rating.

   (B) The total rating of utilization equipment fastened in place, other than luminaires, shall not exceed 50 percent of the branch-circuit amp rating.

(2) 30-amp branch circuits shall be permitted to supply utilization equipment in any occupancy. A rating of any one cord-and-plug-connected utilization equipment shall not exceed 80 percent of the branch-circuit amp rating.

(3) 40- or 50-amp branch circuits shall be permitted to supply cooking appliances that are fastened in place in any occupancy. Larger than 50 amp shall supply only nonlighting outlet loads.

BRANCH CIRCUIT FIXTURE WIRE TAP SIZES

20-amp circuits — 18 AWG, up to a 50 ft run

20-amp circuits — 16 AWG, up to a 100 ft run

20-amp circuits — 14 AWG and larger

30-amp circuits — 14 AWG and larger

40-amp circuits — 12 AWG and larger

50-amp circuits — 12 AWG and larger

MAXIMUM CORD-AND-PLUG-CONNECTED LOAD TO RECEPTACLE		
Circuit Rating (amps)	**Recep. Rating (amps)**	**Maximum Load (amps)**
15 or 20	15	12
20	20	16
30	30	24

RECEPTACLE RATINGS FOR VARIOUS SIZE CIRCUITS	
Circuit Rating (amps)	**Receptacle Rating (amps)**
15	Not over 15
20	15 or 20
30	30
40	40 or 50
50	50

DEMAND FACTORS FOR HOUSEHOLD ELECTRIC CLOTHES DRYERS	
Dryers	**Demand Factor**
1 to 4	100%
5	85%
6	75%
7	65%
8	60%
9	55%
10	50%

DEMAND FACTORS FOR NONDWELLING RECEPTACLE LOADS	
Receptacle Load	**Demand Factor**
First 10 kVA or less	100%
Remainder over 10 kVA	50%

DEMAND FACTORS FOR KITCHEN EQUIPMENT IN NONDWELLING APPLICATIONS

Equipment	Demand Factor
1	100%
2	100%
3	90%
4	80%
5	70%
6 and over	65%

LIGHTING LOAD DEMAND FACTORS

Various Occupancy Types	Portion of Lighting Load Applied in Volt-Amps	Demand Factor in Percent
Dwelling units	First 3000 or less at	100%
	From 3001 to 120,000 at	35%
	Remainder over 120,000 at	25%
Hospitals	First 50,000 or less at	40%
	Remainder over 50,000 at	20%
Hotels and motels, including apartment houses without provision for cooking	First 20,000 or less at	50%
	From 20,001 to 100,000 at	40%
	Remainder over 100,000 at	30%
Warehouses (storage)	First 12,500 or less at	100%
	Remainder over 12,500 at	50%
All others	Total volt-amps of load	100%

LIGHTING LOADS BY OCCUPANCY	
Various Occupancy Types	**Volt-Amps per Square Foot**
Armories	1
Banks	3½
Beauty parlors	3
Churches	1
Clubs	2
Court rooms	2
Dwelling units	3
Garages — commercial	½
Hospitals	2
Hotels and motels, including apartment houses without provision for cooking	2
Industrial commercial buildings	2
Lodge rooms	1½
Office buildings	3½
Restaurants	2
Schools	3
Stores	3
Warehouses	¼
In any of the preceding occupancies except one-family dwellings and individual dwelling units of two-family and multifamily dwellings	
Assembly halls and auditoriums	1
Halls, corridors, closets, stairways	½
Storage spaces	¼

CHAPTER 4
Receptacles, Switches, Interior Wiring and Lighting

RECEPTACLES IN DAMP LOCATIONS

1) All receptacles installed outdoors in locations protected from the weather or in damp locations shall have weatherproof enclosures for such receptacles when the receptacles are covered (attached plug caps not inserted and receptacle covers closed).
2) Installations suitable for wet locations shall also be considered suitable for damp locations.
3) A receptacle shall be considered to be in a location protected from the weather where located under roofed open porches, canopies, etc., and will not be subjected to driving rain or water runoff.
4) All 15 and 20 amp, 125 and 250 volt nonlocking receptacles shall be a listed weather-resistant type and will be identified as 5-15, 5-20, 6-15, and 6-20 by the National Electrical Manufacturers Association (NEMA).

RECEPTACLES IN WET LOCATIONS

1) All 15 and 20 amp, 125 and 250 volt receptacles installed in a wet location shall have an enclosure that is weatherproof whether or not the attachment plug cap is inserted.
2) All 15 and 20 amp, 125 and 250 volt nonlocking receptacles shall be a listed weather-resistant type and will be identified as 5-15, 5-20, 6-15, and 6-20 by NEMA.
3) Receptacles installed in a wet location and subject to routine high-pressure spray washing shall be permitted to have an enclosure that is weatherproof when the attachment plug is removed.

RECEPTACLES IN WET LOCATIONS *(cont.)*

4) A receptacle installed in a wet location where the equipment to be plugged into it is not attended while in use shall have an enclosure that is weatherproof with the attachment plug cap inserted or removed.
5) A receptacle installed in a wet location where the equipment to be plugged into it will be attended while in use shall have an enclosure that is weatherproof when the attachment plug is removed.
6) Receptacles shall not be installed within or directly over a bathtub or shower stall.
7) There shall be proper protection provided for floor receptacles so that floor-cleaning equipment may be operated without damage to such receptacles.
8) The enclosure for a receptacle installed in an outlet box flush-mounted in a finished surface is made weatherproof by installing a weatherproof faceplate that provides a watertight connection between the plate and the finished surface.

TAMPER-RESISTANT RECEPTACLES IN DWELLING UNITS

All 15 and 20 amp, 125 volt receptacles shall be listed as tamper-resistant when they are installed in dwelling units.

ARC-FAULT CIRCUIT-INTERRUPTER PROTECTION

An AFCI receptacle is a device that is intended to provide personal protection from the effects of arc faults by recognizing characteristics unique to arcing and then by de-energizing the circuit when an arc fault is detected.

AFCI RECEPTACLE LOCATIONS

Residential:

Family rooms	Libraries	Rec rooms
Dining rooms	Dens	Closets
Living rooms	Bedrooms	Hallways
Parlors	Sunrooms	Similar rooms/areas

GFCI RECEPTACLE LOCATIONS

Residential:

Bathrooms
Boathouse hoists
Crawl spaces
Garages and accessory buildings
Kitchen countertops
Laundry, utility and wet bar sinks
Outdoors
Swimming pools
Unfinished basements
Whirlpools and hot tubs

Non-Residential:

Bathrooms
Carnivals, circuses and fairs
Commercial garages
Construction sites
Commercial and institutional kitchens
Locker rooms
Marinas
Outdoor public spaces
Rooftops
Swimming pools
Temporary installations
Trailer and RV parks
Vending machines
Whirlpools and hot tubs

For Replacing Equipment in Ungrounded Outlets:

Light fixtures
Receptacles
Toggle switches

GFCI Notes

- A bathroom is defined as an area including a basin with a toilet, tub or shower.
- Reference the *NEC*® for the definition of commercial and institutional kitchens.
- For applications other than light fixtures, receptacles, and switches, GFCI protection is never intended to be a substitute for grounding.

GFCI WIRING DIAGRAMS

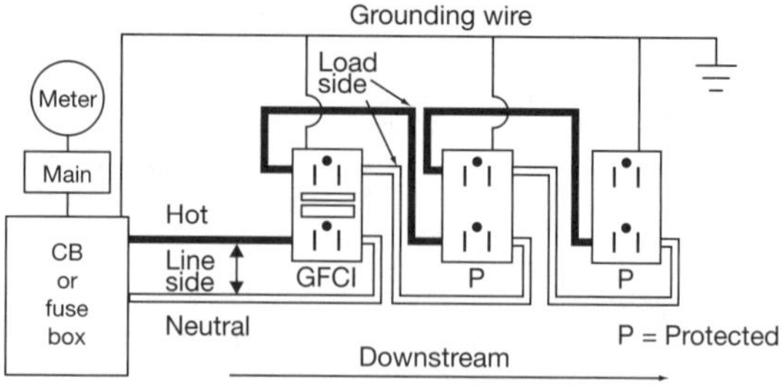

Feed-Thru Installation

To protect the entire branch circuit, the GFCI must be the first receptacle from the circuit breaker or fuse box. Receptacles on the circuit downstream from the GFCI will also be protected.

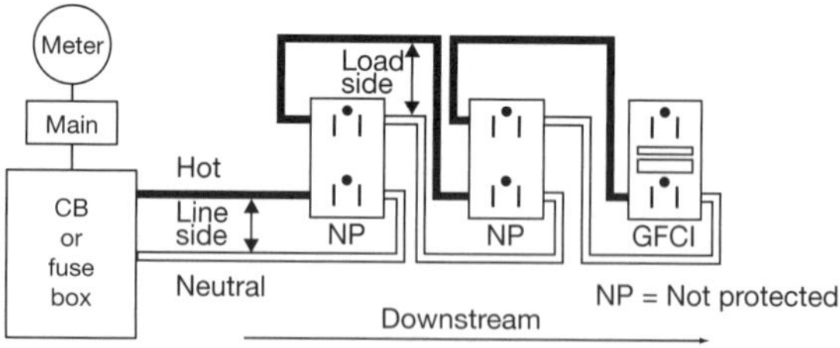

Non-Feed-Thru Installation on a 2-Wire Circuit

Terminal protection can be achieved on a multi-outlet circuit by connecting the hot and neutral line conductors to the corresponding line side terminals of the GFCI. Only the GFCI receptacle will be protected.

WIRING DIAGRAMS FOR NEMA PLUG AND RECEPTACLE CONFIGURATIONS

2-Pole 2-Wire Non-Grounding

125 V

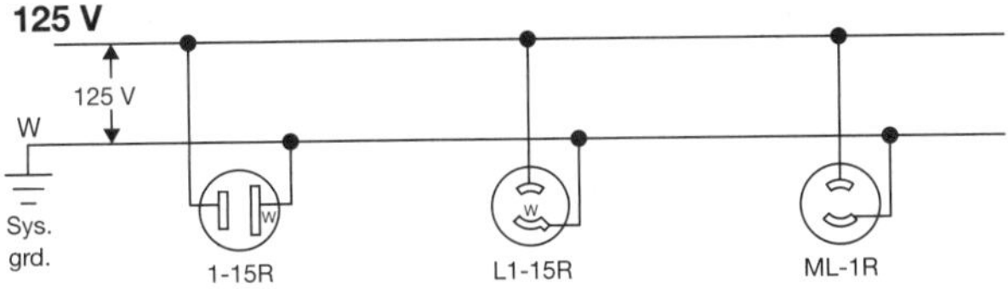

250 V

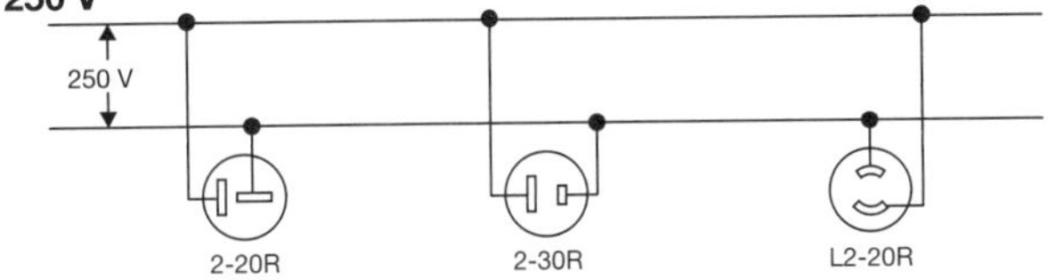

3-Pole 3-Wire Non-Grounding

125/250 V

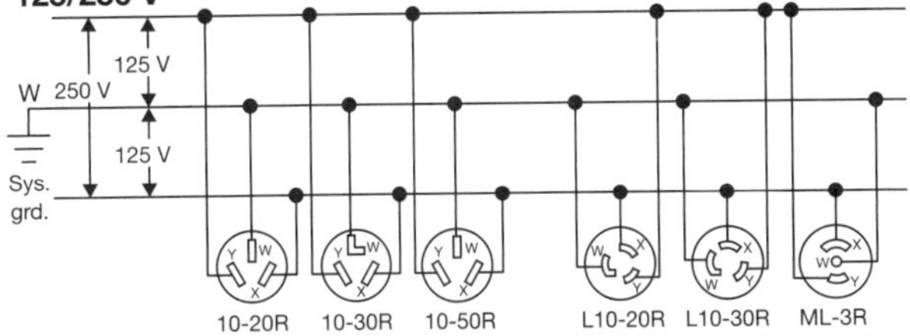

3ϕ 250 V

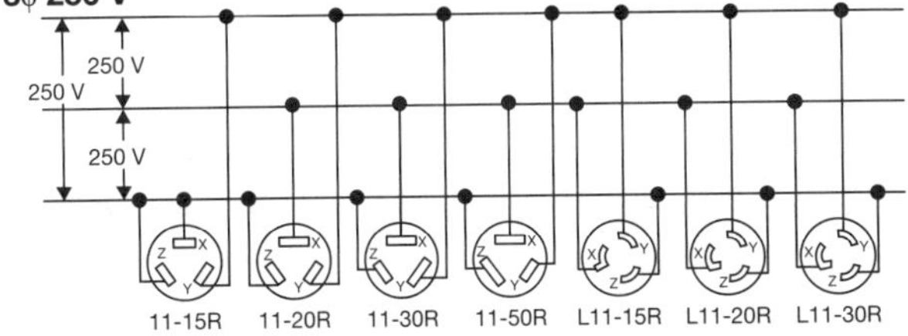

WIRING DIAGRAMS FOR NEMA PLUG AND RECEPTACLE CONFIGURATIONS *(cont.)*

2-Pole 3-Wire Grounding

125 V

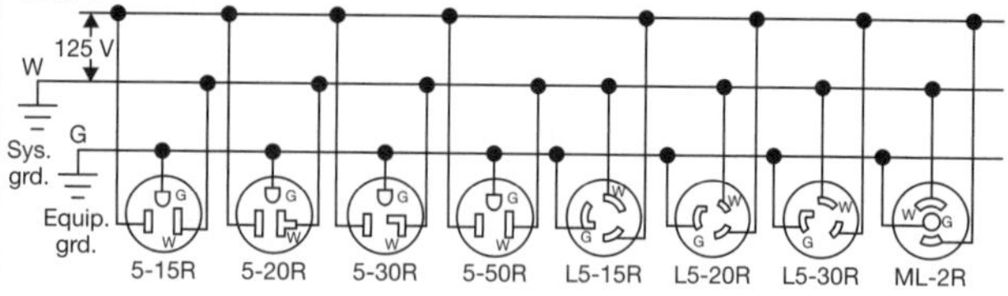

250 V

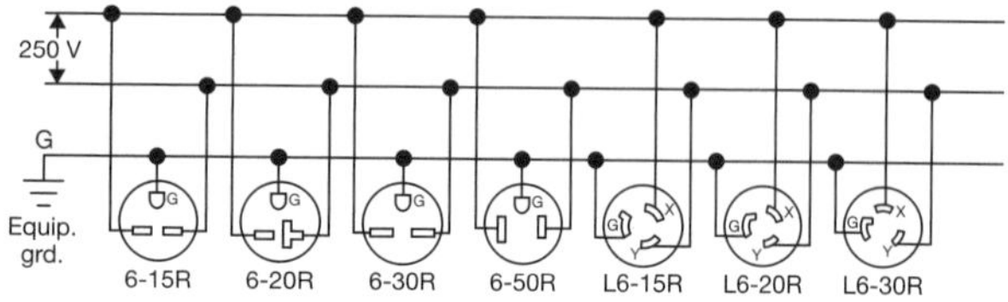

277 VAC

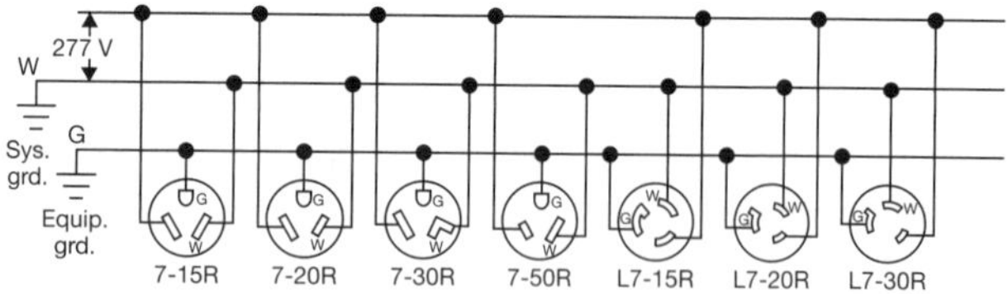

480 VAC

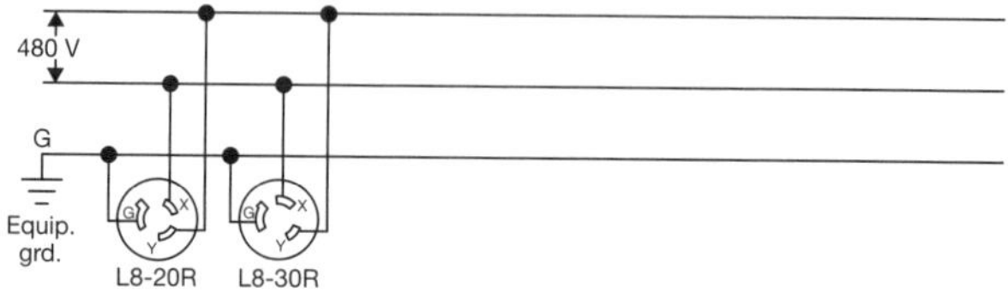

WIRING DIAGRAMS FOR NEMA PLUG AND RECEPTACLE CONFIGURATIONS *(cont.)*

3-Pole 4-Wire Grounding

125 V/250 V

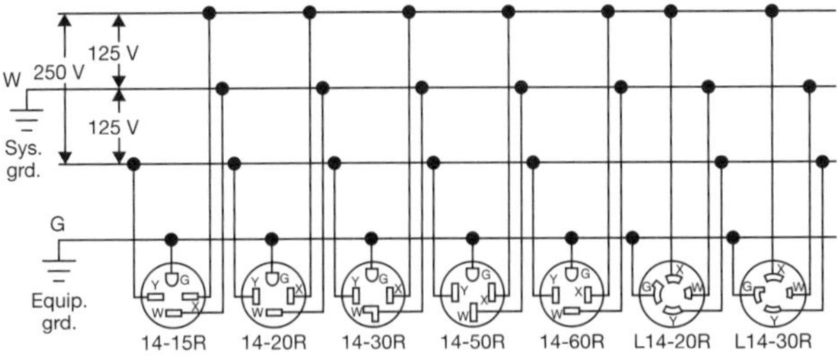

3ϕ 250 V

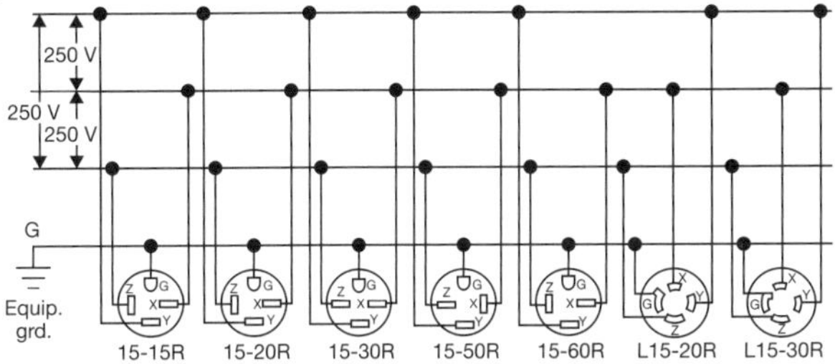

3ϕ 480 V

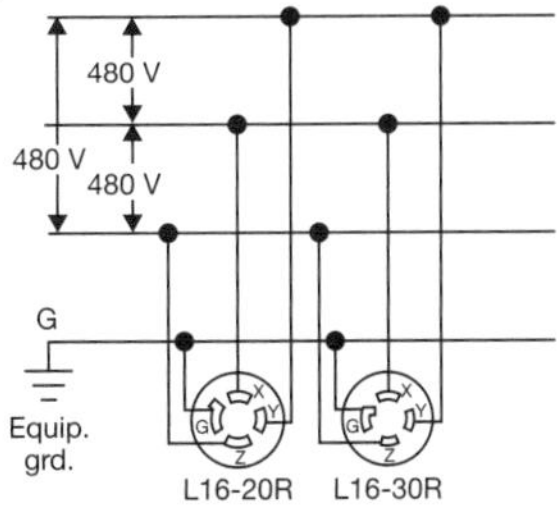

3ϕ 600 V

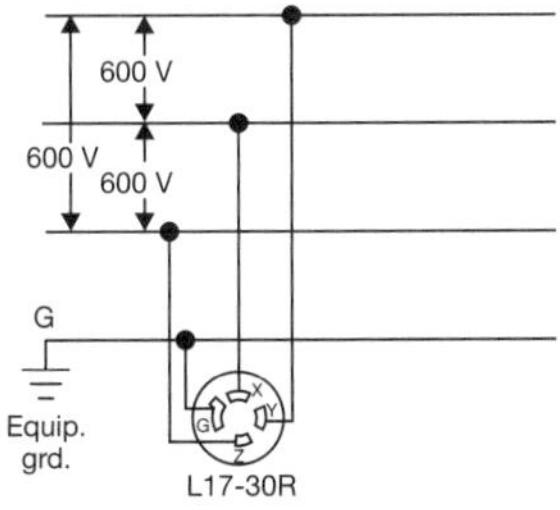

WIRING DIAGRAMS FOR NEMA PLUG AND RECEPTACLE CONFIGURATIONS *(cont.)*

4-Pole 4-Wire Non-Grounding

3ϕ Wye 120 V/208 V

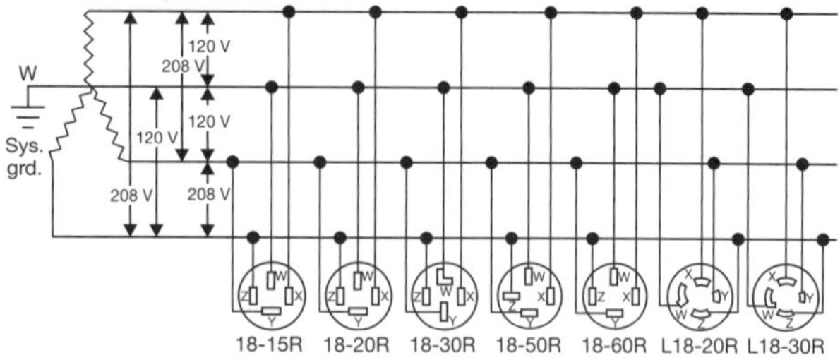

3ϕ Wye 277 V/480 V

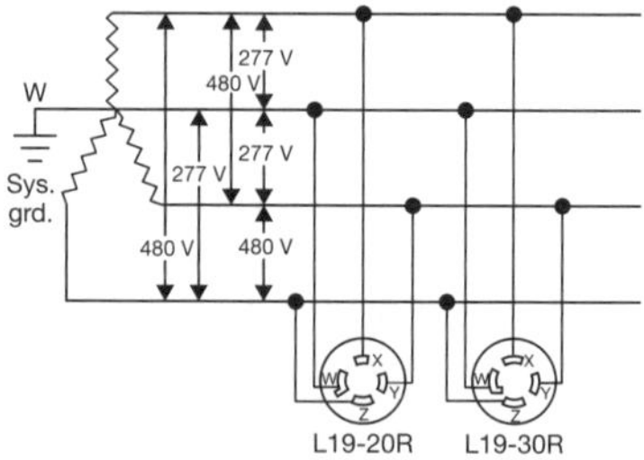

3ϕ Wye 347 V/600 V

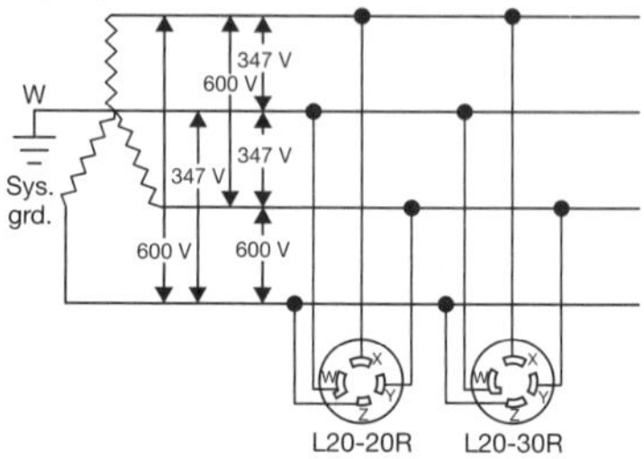

WIRING DIAGRAMS FOR NEMA PLUG AND RECEPTACLE CONFIGURATIONS *(cont.)*

4-Pole 5-Wire Grounding

3ϕ Wye 120 V/208 V

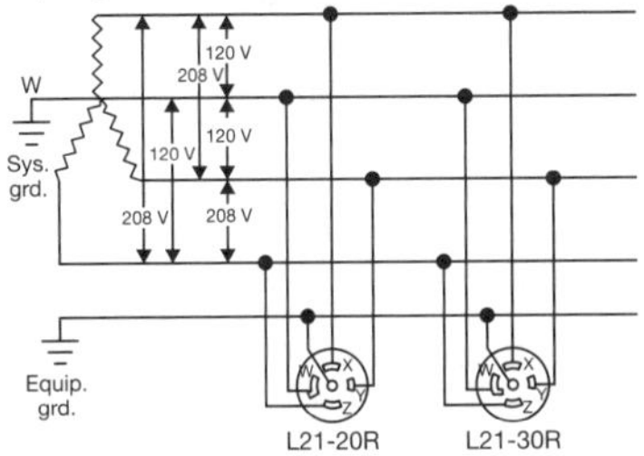

3ϕ Wye 277 V/480 V

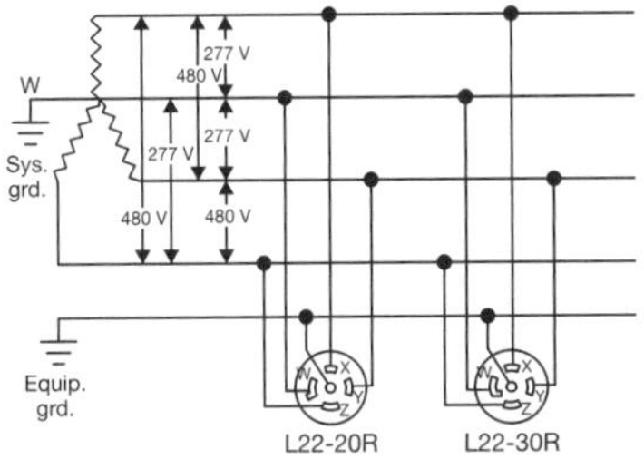

3ϕ Wye 347 V/600 V

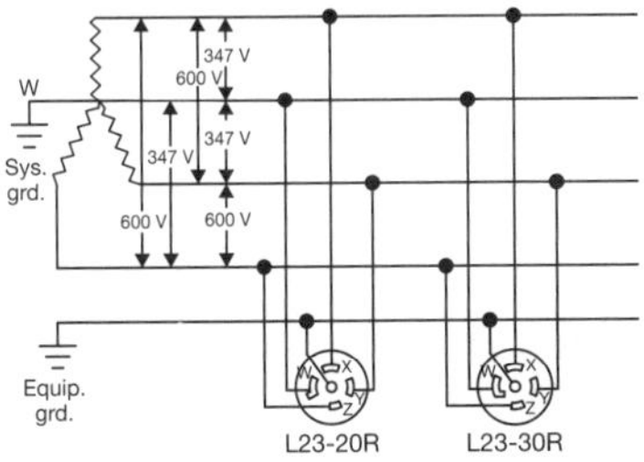

WIRING DIAGRAMS FOR AC SWITCHES

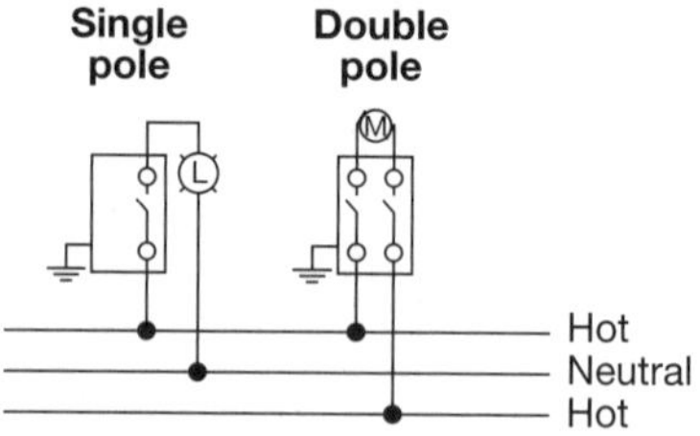

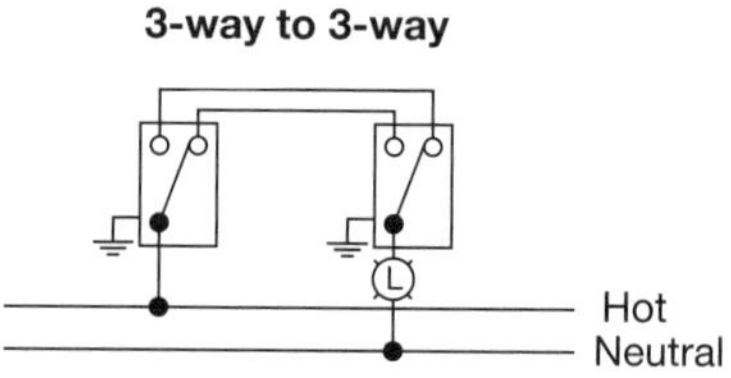

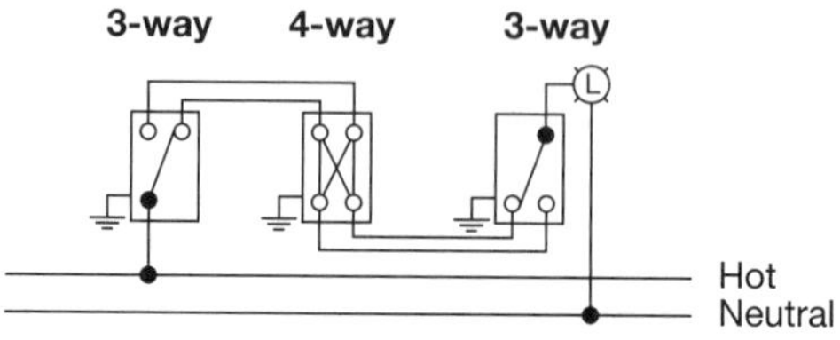

WIRING DIAGRAMS FOR PILOT LIGHT AND LIGHTED TOGGLE SWITCHES

Single Pole Pilot Light Switch
Toggle glows when light is on.

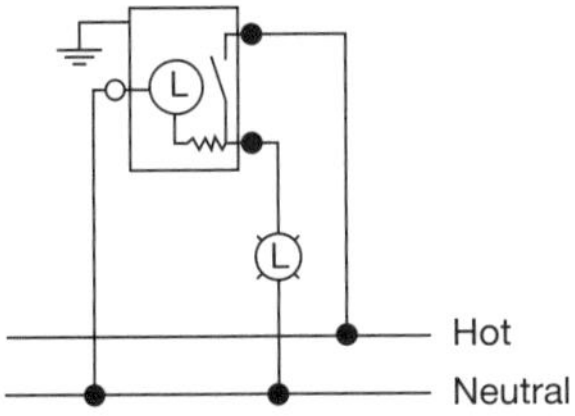

Single Pole Lighted Toggle Switch
Toggle glows when switch is off.

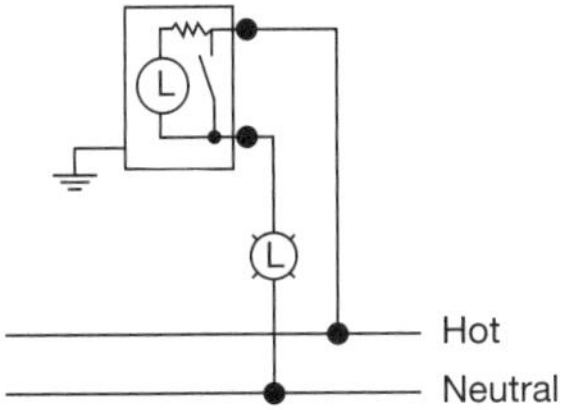

Double Pole Pilot Light Switch
Toggle glows when switch is on.

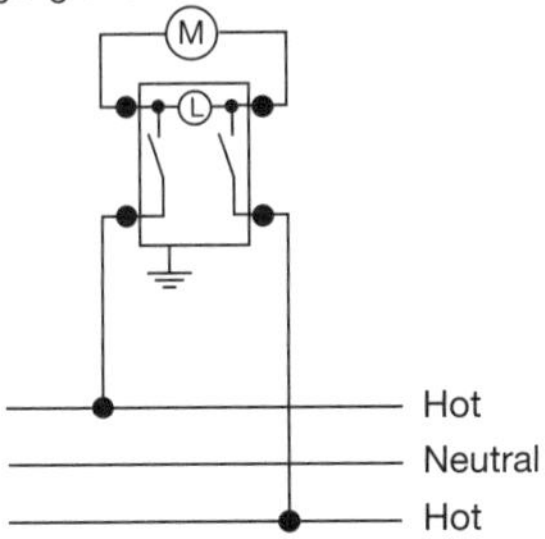

FAMILY ROOM WITH SPLIT-WIRED RECEPTACLES AND SWITCHED CIRCUIT

A split-wired receptacle has the tab between the brass (hot) terminals removed but silver (neutral) terminals remain intact. This provides either a switched circuit or two separate circuits at the same receptacle outlet.

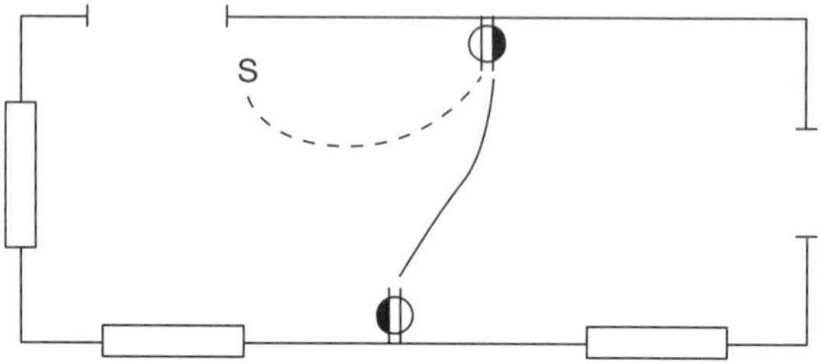

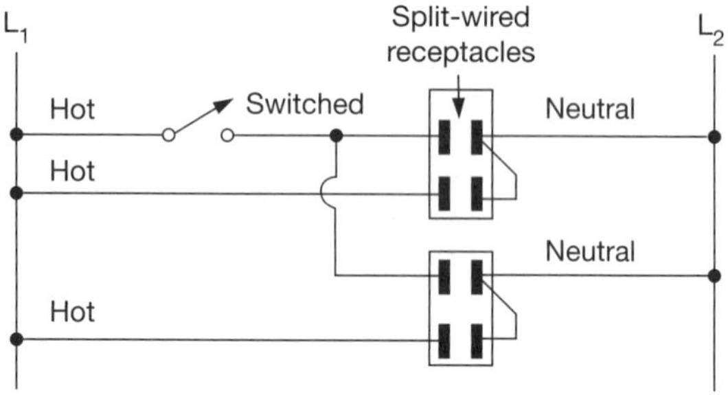

RESIDENTIAL SMOKE ALARM GUIDELINES

New Construction:

- One detector in each bedroom or room with adaptable sleeping furniture.
- One in the hall outside each sleeping area.
- One per floor, including the basement or below grade recreation room or area.
- In basements, locate near the stairs, on the underside of a ceiling joist.
- Wire to a 120 volt power supply circuit and utilize backup batteries.
- Interconnect all detectors with a 3-wire circuit with ground using NM cable.

Existing Construction:

- Add smoke detectors to existing houses when a permit is pulled for work inside the house.
- Install detectors in the same areas that are required for new construction.
- Battery powered units may be used where you are unable to wire to the location of the smoke detector.

Do Not Install:

- Within 4" of the junction between wall and ceiling (dead air space).
- Within 3' of doorway to bathrooms and kitchens.
- Within 3' of air supply vents and the tips of paddle fans.
- Within 20' of a cooking appliance (unless the alarm has a push or silence button, or if photoelectric type).
- In dusty attics that get too hot or cold.
- Where protected by a GFCI.

RESIDENTIAL AREA REQUIREMENTS FOR SMOKE ALARM

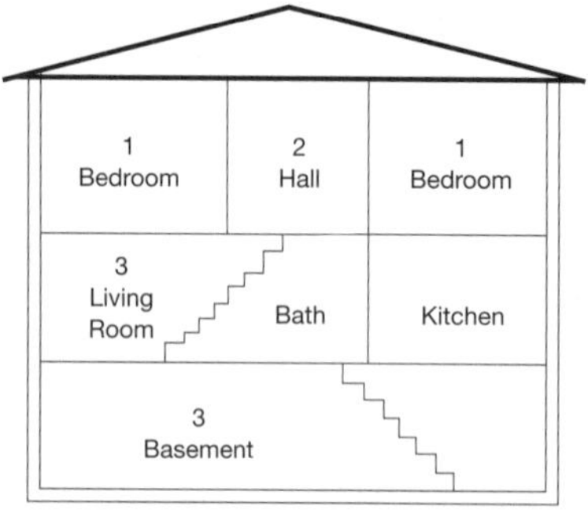

Number	Area Requirement
1	Smoke alarms must be located in each sleeping room.
2	Smoke alarms must be located outside of each sleeping area.
3	Smoke alarms must be located on each level of the dwelling unit including basements, but NOT including crawl spaces or uninhabitable attics.

SMOKE ALARM WIRING REQUIREMENTS

- 120 volt power supply from AFCI protected circuit.
- Smoke alarms must be interconnected so that actuation of one alarm will activate alarms on all smoke alarms.
- Must receive primary power from building wiring (permanent) with battery backup.
- No disconnecting switch other than a circuit breaker.

FAMILY ROOM AND BEDROOM RECEPTACLE OUTLET SPACING

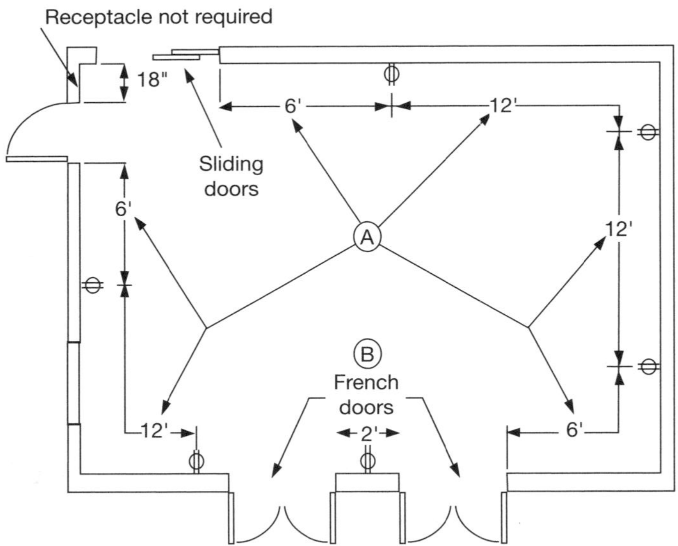

(A) No point along floor line shall be more than 6' from a receptacle outlet.

(B) All wall spaces 2' or larger shall have a receptacle outlet.

FAMILY ROOM WIRING DETAILS

- Receptacle outlets must be spaced so that no point along the wall is farther than 6' from the nearest outlet. A fireplace, doorway, or similar opening is not required to be counted as wall space, but receptacles must be located within 6' of each side of these spaces.
- Wall space 2' or greater must have a receptacle outlet.
- Fixed panels such as the portion of a sliding glass door that is not movable must be considered wall space.
- Switched lighting outlet is required. Lighting fixture or switched plug is acceptable.
- GFCI protection required for all 120 volt, 15 or 20 amp receptacles located within 6' of a wet bar sink.
- If wall space is not available, receptacles must be installed on the floor. Receptacles installed on the floor must be within 18" of the wall to be counted as a required receptacle.
- All 15 and 20 amp, 120 volt receptacle outlets must be listed as tamper-resistant.
- Arc-fault protection is also required for all outlets including all lighting fixtures and receptacles for family rooms including wet-bar areas, bonus rooms or living rooms.

WETBAR AND TRACK LIGHTING WIRING DETAILS

- GFCI protection required for all 120 volt, 15 or 20 amp receptacles located within 6' of a wetbar sink and there are no exceptions for single outlet or dedicated appliances.
- GFCI protection is NOT required for receptacles located 6' or further from the outside edge of a wetbar sink, but they must be AFCI protected and listed tamper-resistant.
- Space occupied by a countertop must be counted as wall space and satisfy the 6' requirement for placing receptacles along the wall.
- Track lighting is not allowed less than 5' above the floor unless protected from possible damage.
- Track lighting must be hard wired and must be supported at a minimum of two points every 4'.

BEDROOM WIRING DETAILS

- Receptacle outlets must be spaced so that no point along the wall is farther than 6' from the nearest outlet.
- Wall space 2' or greater must have a receptacle outlet.
- Switched lighting outlet is required. Lighting fixture or switched plug is acceptable.
- Smoke detector required inside all sleeping rooms.
- Arc-fault protection required for all outlets which includes all lighting fixtures, smoke detectors, and receptacles.
- All 15 and 20 amp, 120 volt receptacle outlets must be listed tamper-resistant.

CLOTHES CLOSET WIRING DETAILS

- Open bulb incandescent lighting fixtures are not allowed.
- Storage space is considered 12" (minimum) from side and rear walls or the shelf width, whichever is greater.
- Recessed and enclosed incandescent and fluorescent lights must be 6" from nearest storage space and 18" from side and rear walls.
- Surface mounted incandescent lights must be 12" from nearest storage space and 24" from side and rear walls.
- Surface mounted fluorescent lights must be 6" from nearest storage space and 18" from side and rear walls.
- All 15 and 20 amp, 120 volt outlets must be AFCI protected and listed tamper-resistant.
- Lighting and receptacle outlets are not required unless it is a walk in closet.

HALLWAY AND STAIRWELL WIRING DETAILS

- Wall switch controlled lighting outlet required.
- Where there are 6 or more risers on the stairs, a 3 way or 4 way wall switch to control the lighting outlet is required at each level.
- Hallways 10' or longer must have a receptacle outlet.
- Smoke detectors must be located outside of each sleeping area.
- All 15 and 20 amp, 120 volt outlets must be AFCI protected and listed tamper-resistant.

LAUNDRY ROOM WIRING DETAILS

- Switched lighting outlet required.
- Dedicated 20 amp circuit required for laundry receptacle outlets. No other devices allowed. At least 1 receptacle outlet required for laundry purposes.
- GFCI protection required for all 120 volt, 15 or 20 amp receptacles located within 6' of a utility sink. (No exception for dedicated appliance such as a washer).
- 4 wire, 30 amp circuit required for electric dryers.
- All 15 and 20 amp, 120 volt receptacle outlets must be listed tamper resistant.
- Clothes dryers come from the factory with a bonding strap joining the ground and neutral terminals. This strap must be removed to comply with the code. Follow manufacturer's instructions to make cord connections.

KITCHEN RECEPTACLES

All receptacles must be on small appliance branch circuits. GFCI protection is required for all receptacles serving kitchen countertops and countertop surfaces.

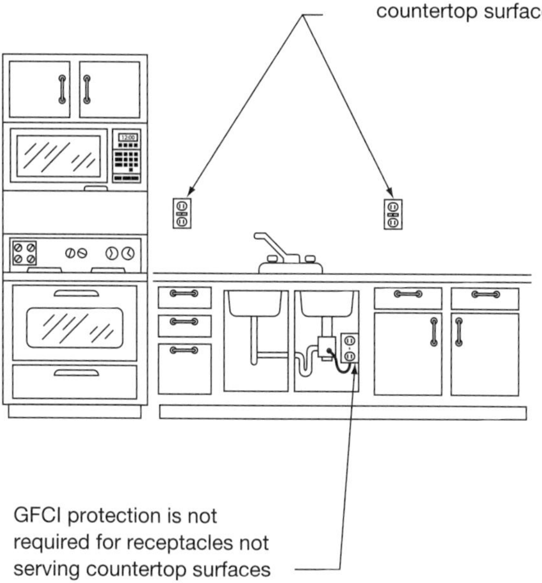

GFCI protection is not required for receptacles not serving countertop surfaces or for receptacles serving dedicated appliances such as a garbage disposal.

COUNTERTOP RECEPTACLE OUTLET SPACING

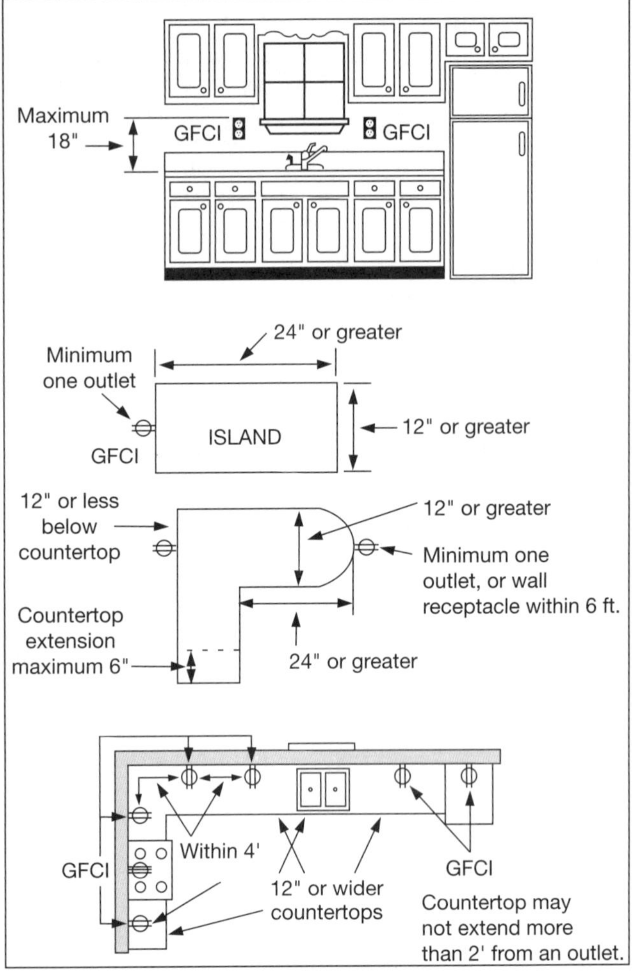

KITCHEN WIRING DETAILS

- No point along the countertop is further than 2' from a receptacle and counter spaces longer than 1' require a receptacle.
- One wall switch controlled lighting fixture required.
- If spacing behind the sink is less than 12", one receptacle is required within 2' of each side.
- If spacing behind the sink is 12" or greater, this area must be counted as counter space.
- Counter spaces divided by range tops, refrigerators, or sinks must be considered separate spaces.
- An island or peninsula space requires a receptacle only if larger than 24" × 12". If a sink or cooktop is installed on the island, and the width of the counter behind the appliance is less than 12", a receptacle must be installed on both sides. (A countertop wall receptacle within 6 ft. is acceptable).
- If spacing behind cooktop is less than 18 in. one receptacle is required within 2' of each side.
- If spacing behind cooktop is 18" or greater, this area must be counted as wall space.
- No point along the wall may be further than 6' from a receptacle, if not counter space.
- All 15 and 20 amp, 120 volt receptacle outlets in kitchens must be listed tamper-resistant.
- All 15 and 20 amp, 120 volt receptacle outlets in dining rooms and eating areas must be AFCI protected and listed tamper-resistant.
- Countertop surface receptacles must be GFCI protected.
- All receptacles in kitchen area must be on a 20 amp circuit.
- Minimum of two 20 amp circuits required to supply kitchen counter top receptacles.

KITCHEN WIRING DETAILS *(cont.)*

- Receptacle for refrigerator should be supplied from a dedicated 15 or 20 amp circuit.
- Garbage disposal flex cord (use 3 wire type) can be no shorter than 18" and no longer than 36" and can be hard wired.
- Disposal receptacle must be located in the space occupied by the dishwasher or adjacent cabinet.
- Switch is allowed for disposal disconnect if not cord-and-plug connected.
- Length of dishwasher cord (use 3 wire type) must extend 3'–4' from the rear of the appliance. Dishwasher is normally hard wired.
- Receptacles for microwaves should be mounted in the cabinet directly above the appliance. GFCI protection is not required. Microwaves are normally supplied by a dedicated 20 amp circuit, 12/2 with ground, copper, nonmetallic cable.
- Ranges rated 8¾ kW or more must be supplied by a minimum 40 amp circuit.
- Receptacles to be mounted no more than 18" above the countertop.
- 4 wire circuit required for cord-and-plug connected free standing ranges which are typically wired with 50 amp circuit, 6/3 with ground, copper, nonmetallic cable.
- Range hoods and built-in microwaves may be cord-and-plug connected if the receptacle is accessible and supplied from an individual circuit.

BATHROOM RECEPTACLES AND SPACING

At least one receptacle outlet shall be installed within 36" of outside edge of each basin and on a wall adjacent to basin location.

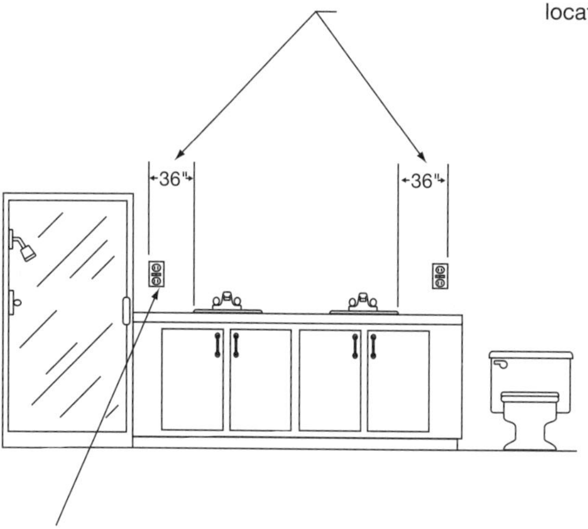

All GFCI-protected receptacles shall be supplied by at least one 20 amp circuit with no other type of outlets on the circuit.

All bathroom receptacles in dwelling units are required to be GFCI-protected.

BATHROOM WIRING DETAILS

- One wall switch controlled lighting outlet required.
- All receptacles must be GFCI protected and wired on a 20 amp circuit and at least one receptacle must be installed within 3' of each sink.
- No other devices such as lights or vent fans are allowed on the receptacle circuit, unless the dedicated 20 amp circuit supplies only one bathroom.
- Different bathroom receptacles may be served from one 20 amp circuit, but only those receptacles, not any other devices.
- No part of a lighting track, hanging lighting fixture or ceiling fan allowed within a 3' horizontal by 8' vertical zone above the threshhold of a shower or the rim of a bathtub. Recessed or surface mounted fixtures are allowed, but must be listed for damp locations.
- Receptacles or switches are not allowed within or directly over a tub or shower space.
- Access is required to all electrical equipment for hydromassage bathtubs.
- Hydromassage bathtub electrical equipment must be GFCI protected and wired on a dedicated 20 amp circuit.
- Metal piping systems must be bonded to pump motor with #8 solid copper wire.
- All 15 and 20 amp, 120 volt receptacle outlets must be listed tamper resistant.
- Vent fans installed directly above showers must be GFCI protected.

CRAWL SPACE AND INSIDE EQUIPMENT WIRING DETAILS

- Receptacle required if heating or air conditioning equipment is present. Receptacle must be located on the same level and within 25' of the equipment.
- Receptacles must be GFCI protected.
- Lighting outlet required if space is used for storage or contains equipment that may need servicing and must be located at or near the equipment (HVAC, etc.).
- Switch for lighting outlet must be located at the point of entrance to the storage or equipment space.
- Disconnecting means for HVAC equipment must be located within sight and readily accessible from the equipment. The disconnect may not be mounted on the panels designed to allow access to the equipment.
- All 15 and 20 amp, 120 volt receptacle outlets in dwellings must be listed tamper-resistant.
- All 15 and 20 amp, 120 and 240 volt nonlocking receptacles in wet or damp locations must be listed weather-resistant type.
- Cables containing wires smaller than two #6 or three #8 are not allowed to be secured directly to the bottom of the ceiling joists. These wires must be routed through bored holes, installed on running boards or through conduit raceways.
- The access must be sized large enough to remove the largest piece of equipment out of the space.
- The passageway to the equipment must be floored and not less than 24" wide or wider to allow equipment installation or removal.
- A 30" × 30" level space must be provided in front of the equipment for servicing.

UNFINISHED BASEMENT WIRING DETAILS

- Wall switch controlled lighting outlet is required.
- Switched lighting outlet required on exterior to illuminate all personnel entrances.
- At least one receptacle outlet is required in unfinished basements.
- Where a portion of the basement is finished into a habitable room, one receptacle outlet is required in each unfinished portion.
- All 15 and 20 amp, 120 volt receptacle outlets must be listed tamper-resistant.
- GFCI protection is required for all 15 and 20 amp, 120 volt receptacles.
- GFCI protection is required for all 15 and 20 amp, 120 volt receptacles that supply sump pumps.
- GFCI protection required if receptacles are not readily accessible such as ceiling receptacles for sump pump alarms.
- GFCI protection required for receptacles that are dedicated for appliances that are not easily moved such as a refrigerator, freezer or washer.
- 15 and 20 amp, 120 volt receptacles must be GFCI protected within 6' of a utility sink.
- Exception: GFCI protection not required for a receptacle supplying a permanently installed burglar or fire alarm system.
- Smoke alarm must be installed for the basement level.
- NM cable must closely follow the surface of the building or be protected from damage. (Must protect horizontal runs of exposed cable.)
- NM cable used on the wall of an unfinished basement is permitted to be installed in listed conduit. (Must use non-metallic bushing at the point where cable enters.)
- Cables containing wires smaller than two #6 or three #8 are not allowed to be secured directly to the bottom of the ceiling joists. (Must be routed through bored holes, installed on running boards or in conduit raceways.)

GARAGE AND SHOP RECEPTACLES

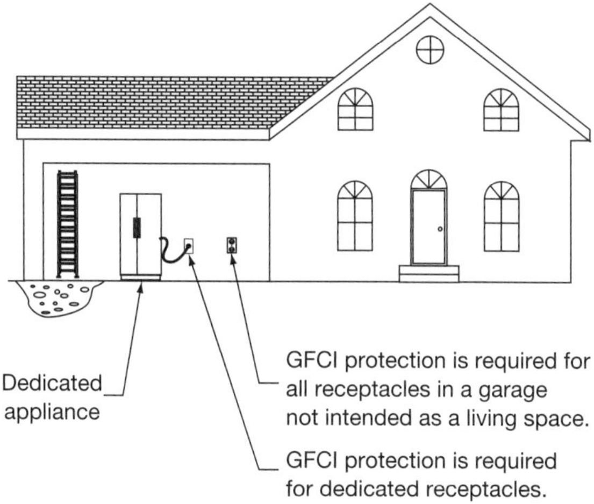

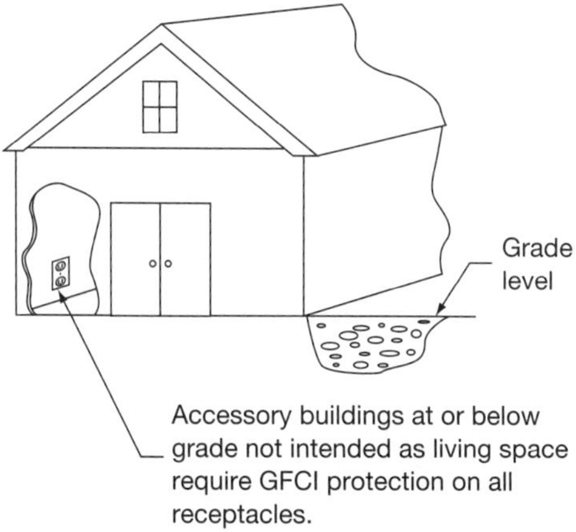

GARAGE AND SHOP WIRING DETAILS

- Wall switch controlled lighting outlet is required.
- Switched lighting outlet, required on exterior to illuminate all personnel entrances, but is not required for vehicle doors, such as overhead garage doors.
- At least one receptacle outlet is required in attached garages, detached garages and shops.
- GFCI protection is required for all 15 and 20 amp, 120 volt receptacles.
- GFCI protection required if receptacles are not readily accessible such as a ceiling receptacle for a garage door opener.
- GFCI protection required for receptacles that are dedicated for appliances that are not easily moved such as a refrigerator, freezer or a washer.
- 15 and 20 amp, 120 volt receptacles must be GFCI protected within 6' of a utility sink.
- All 15 and 20 amp, 120 volt receptacle outlets must be listed tamper-resistant.
- Some jurisdictions require smoke detectors and/or carbon monoxide detectors for garages. Always check with your local building codes or the electrical inspection department.

STORAGE AREA WIRING DETAILS

- Switched lighting outlet is required and must be located at the point of entrance to the space.
- All 15 and 20 amp, 120 volt receptacle outlets must be listed tamper-resistant.
- Receptacle outlet required for HVAC equipment.
- Smoke detectors not required in uninhabitable attics.

OUTDOOR RECEPTACLES

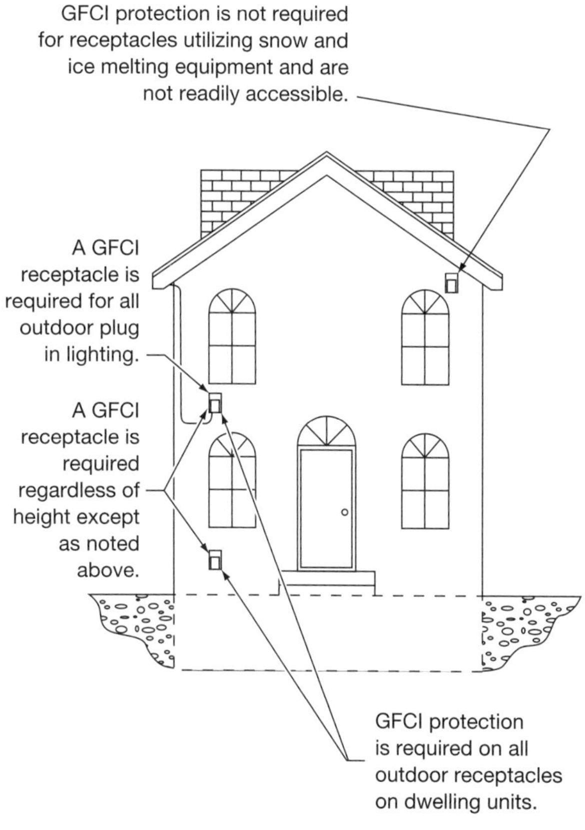

OUTDOOR WIRING DETAILS

- Receptacle outlet required at the front and rear located no higher than 6' 6" above grade.
- All 15 and 20 amp, 120 volt receptacles must be GFCI protected.
- Receptacles installed in damp locations must have a weather resistant cover.
- Receptacles installed in wet locations must have a cover that is weatherproof whether or not a cord is plugged in.
- Switched lighting outlet required to illuminate all personnel entrances.
- All 15 and 20 amp, 120 volt receptacle outlets must be listed tamper-resistant.
- All 15 and 20 amp, 120 and 240 volt nonlocking receptacles in wet or damp locations must be listed weather-resistant-type.
- HVAC equipment must have receptacle on the same level and located within 25'.
- Disconnecting means for HVAC equipment must be located within sight and readily accessible from the equipment.
- Must maintain clear working space at all electrical equipment that may require servicing while energized.
- Conduit installed outside in a wet location must be listed for use outdoors.
- Conductors installed outdoors must be listed for use in wet locations.

RECOMMENDED LIGHT LEVELS			
Interior		**Exterior**	
Area	**fc**	**Area**	**fc**
Assembly		**Airports**	
Rough	30	Terminal	10
Medium	100	Loading	2
Banks		**Buildings**	
Lobby	50	Light surface	15
Tellers	150	Dark surface	50
Hospital/Med.		**Construction**	
Dental	1000	General	10
Operating	2500	Excavation	2
Machine Shop		**Parking Areas**	
Rough	50	Industrial	2
Medium	100	Shopping	5
Offices		**Loading Areas**	
Regular	100	Pier	20
Detailed work	200	Trucking	30
Printing		**Service Station**	
Proofreading	150	Pumps	25
Color inspecting	200	Service	5

LAMP ADVANTAGES AND DISADVANTAGES

Lamp	Advantages	Disadvantages
Incandescent, tungsten-halogen	Low initial cost Simple construction No ballast required Available in many shapes and sizes Requires no warm-up or restart time Inexpensively dimmed Simple maintenance	Low electrical efficiency High operating temperature Short life Bright light source in small space Does not allow large distribution of light
Fluorescent	Available in many shapes and sizes Moderate cost Good electrical efficiency Long life Low shadowing Low operating temperature Short turn-ON delay	Not suited for high-level light in small, highly-concentrated applications Requires ballast Higher initial cost than incandescent lamps Light output and color affected by ambient temperature Expensive to dim
Low-pressure sodium, mercury-vapor, metal-halide, high-pressure	Good electrical efficiency Long life High light output Slightly affected by ambient temperature	May cause color distortion Long start and restart time High initial cost High replacement cost Requires ballast Expensive or not possible to dim Problem starting in cold weather High-socket voltage required

LAMP RATINGS

Lamp (W = Watts)	Initial Lumen	Mean Lumen
40 W standard incandescent	480	N/A
100 W standard incandescent	1750	N/A
40 W standard fluorescent	3400	3100
100 W tungsten-halogen	1800	1675
100 W mercury-vapor	4000	3000
250 W mercury-vapor	12,000	9800
100 W high-pressure sodium	9500	8500
250 W high-pressure sodium	30,000	27,000
250 W metal-halide	20,000	17,000

LUMENS, LIFE, AND EFFICACY FOR VARIOUS LAMPS

Lamp Type (Watts)	Initial Lumens	Life (hours)	Efficacy (lumens/ watts)
Incandescent, standard inside frosted			
25	235	1000	9
40	480	1500	12
60	840	1000	14
75	1210	850	16
100	1670	750	17
150	2850	750	19
200	3900	750	19
300	6300	1000	21
500	10750	1000	21
1000	23100	1000	23
Incandescent PAR-38			
150 Spot	1100	2000	7
150 Flood	1350	2000	9
Quartz incandescent			
500	10550	2000	21
1000	21400	2000	21
1500	35800	2000	24
40	3150	20000	78
40	2200	20000	55
60	4300	12000	41
60	3050	12000	50
85	2850	20000	81
85	2000	20000	57
Fluorescent, U-line			
40	2850	12000	71
40	2020	12000	50
Fluorescent, instant start			
60	5600	12000	93
60	4000	12000	66
75	6300	12000	84
75	4500	12000	60
95	8500	12000	89
95	6100	12000	64
110	9200	12000	83
110	6550	12000	59
180	12300	10000	68
215	14500	10000	67
215	13600	10000	63

LUMENS, LIFE, AND EFFICACY FOR VARIOUS LAMPS *(cont.)*			
Lamp Type (Watts)	**Initial Lumens**	**Life (hours)**	**Efficacy (lumens/ watts)**
Mercury-vapor			
40	1140	24000	28
50	1575	24000	31
75	2800	24000	37
100	4300	24000	43
175	8500	24000	48
175	7900	24000	45
250	13000	24000	52
250	12100	24000	48
400	23000	24000	57
400	21000	24000	52
700	43000	24000	61
1000	63000	24000	63
Metal-halide			
175	14000	7500	80
250	20500	7500	82
400	34000	15000	85
1000	105000	10000	105
High-pressure sodium			
50	3300	20000	66
70	5800	20000	82
100	9500	20000	95
150	16000	24000	106
150	12000	12000	80
200	22000	24000	110
215	19000	12000	88
250	27500	24000	110
400	50000	24000	125
1000	140000	24000	140
Low-pressure sodium			
65	4800	18000	137
65	8000	18000	145
90	13500	18000	150
135	22500	18000	166
180	33000	18000	183

LIGHT SOURCE CHARACTERISTICS

Type of Characteristic	Incandescent, Including Tungsten	Fluorescent	High-Intensity Discharge: Mercury-Vapor (Self-Ballasted)	High-Intensity Discharge: Metal-Halide	High-Intensity Discharge: High-Pressure Sodium (Improved Color)	Low-Pressure Sodium
Wattages (lamp only)	15–1500	15–219	40–1000	175–1000	70–1000	35–180
Life[a] (hr)	750–12,000	7500–24,000	16,000–15,000	1500–15,000	24,000 (10,000)	18,000
Efficacy[a] (lumens/W) lamp only	15–25	55–100	50–60 (20–25)	80–100	75–140 (67–112)	Up to 180
Lumen maintenance	Fair to excellent	Fair to excellent	Very good (good)	Good	Excellent	Excellent
Color rendition	Excellent	Good to excellent	Poor to excellent	Very good	Fair	Good
Light direction control	Very good to excellent	Fair	Very good	Very good	Very good	Fair
Source size	Compact	Extended	Compact	Compact	Compact	Extended
Relight time	Immediate	Immediate	3–10 min	10–20 min	Less than 1 min	Immediate
Comparative fixture cost	Low: simple fixtures	Moderate	Higher than incandescent and fluorescent	Generally higher than mercury	High	High
Comparative operating cost	High: short life and low efficiency	Lower than incandescent	Lower than incandescent	Lower than mercury	Lowest of HID types	Low
Auxiliary equipment needed	Not needed	Needed: medium cost	Needed: high cost	Needed: high cost	Needed: high cost	Needed: high cost

[a]Life and efficacy ratings subject to revision. Check manufacturers' data for latest information.

COMMON FLUORESCENT TYPES

Type	Watts	Length	Size	Base
T-5	4 – 13	6 – 21	⅝	Miniature Bi-pin
T-5	13	21	⅝	Miniature Bi-pin
T-8	15 – 30	18 – 36	1	Medium Bi-pin
T-9	20 – 40	6 – 16	1⅛	Four-pin
T-10	40	48	1¼	Medium Bi-pin
T-12	14 – 75	15 – 96	1½	Medium Bi-pin
T-12	60 – 75	96	1½	Single-Pin

Note: All lengths and sizes are in inches.

COMMON SHAPE IDENTIFIERS

C – Circular lamp. The letter "C" typically appears in the prefix, immediately after the "F."

T – Tubular lamp, omitted from the F40 lamp code for rapid-start lamps.

J – Used in conjunction with a "T" designator, "J" denotes a lamp with a clear outer glass jacket over the regular "T" envelope, which acts as a thermal insulator and permits operation of the lamps at sub-zero temperatures.

BX, **PL**, and **PTT**
– These identifiers denote a family of single ended compact lamps.

U – A lamp shaped like a letter "U."

CHAPTER 5
Grounding and Bonding

RESISTIVITIES OF DIFFERENT METALS			
Metal Type	**Ohms/Ft**	**Metal Type**	**Ohms/Ft**
Brass	43	Aluminum	17
Copper	11	Monel	253
Silver	10	Nichrome	600
Gold	15	Nickel	947
Iron (Pure)	60	Tantalum	93
Magnesium	276	Tin	69
Manganin	265	Tungsten	34

Note: The values above represent the resistance offered by a wire one foot long with a diameter of 1 mil at an ambient temp. of 30°C. Resistance varies with temperature change.

RESISTIVITIES OF DIFFERENT SOILS			
	Resistivity in OHMS per cubic meter		
Soil Type	**Average**	**Min.**	**Max.**
Fills – ashes, cinders, brine wastes	2,370	590	7,000
Clay, shale, gumbo, loam	4,060	340	16,300
Same – with varying proportions of sand and gravel	15,800	1,020	135,000
Gravel, sand, stones, with little clay or loam	94,000	59,000	458,000

BASIC GROUNDED CONDUCTOR RULES

Circuit breakers or switches shall not disconnect the grounded conductor of a circuit.

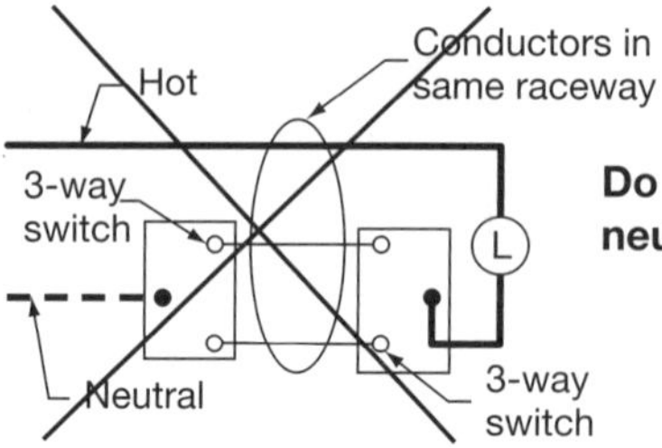

Do not switch the neutral conductor.

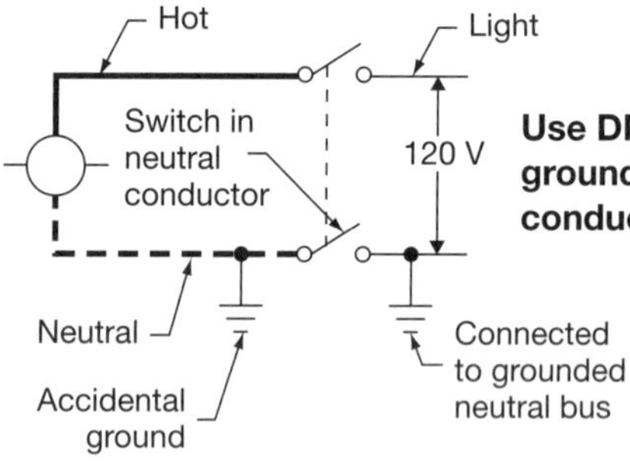

Use DP switch in grounded neutral conductor.

Exception: A circuit breaker or switch may disconnect grounded circuit conductor if all circuit conductors are disconnected at the same time.

BASIC GROUNDED CONDUCTOR RULES *(cont.)*

Circuit breakers or switches shall not disconnect the grounded conductor of a circuit.

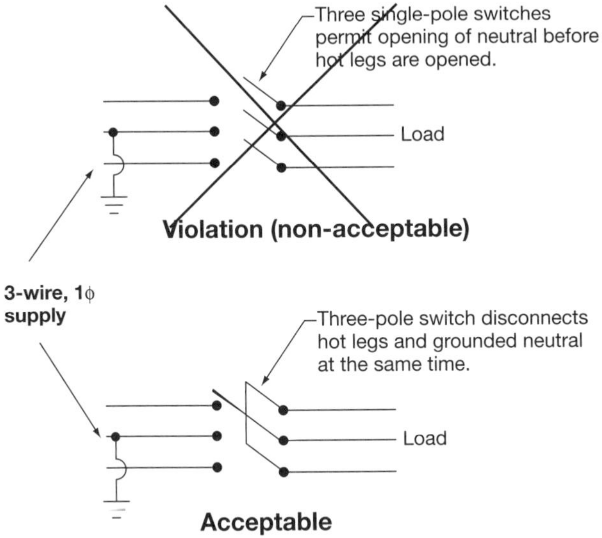

Exception: A circuit breaker or switch may disconnect the grounded circuit conductor if it cannot be disconnected until all other ungrounded conductors have been disconnected.

GROUNDED CONDUCTOR — NEUTRAL

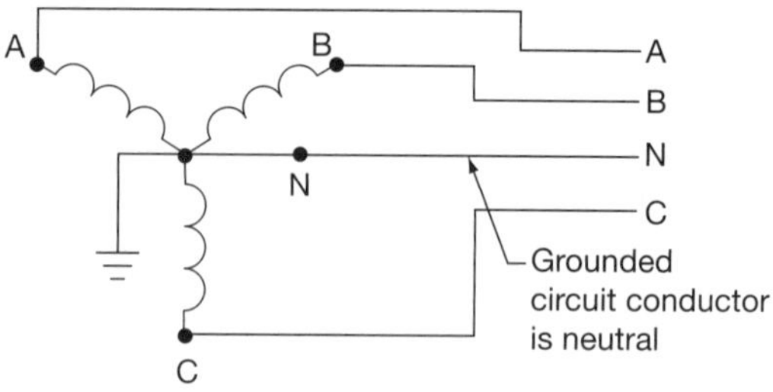

3φ, 4-Wire Wye Grounded System

GROUNDED CONDUCTOR — NOT NEUTRAL

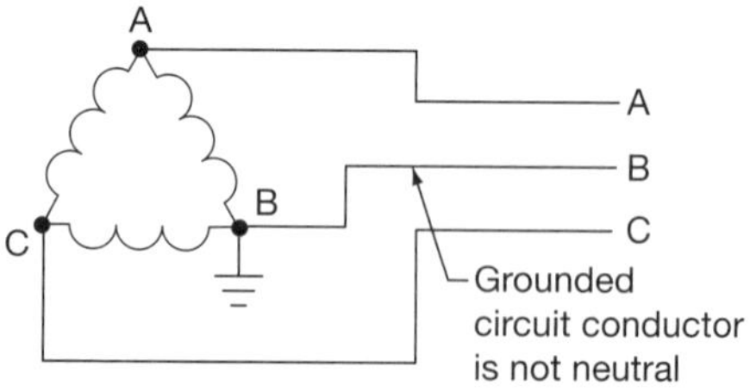

3φ, 3-Wire Corner-Grounded Delta System

GROUNDING DIFFERENT TYPES OF CIRCUITS

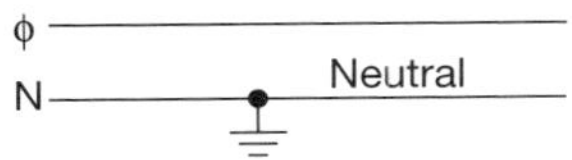

Single-Phase, Two-Wire

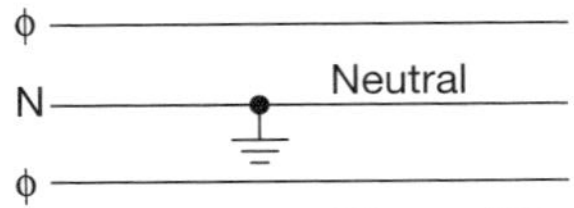

Single-Phase, Three-Wire

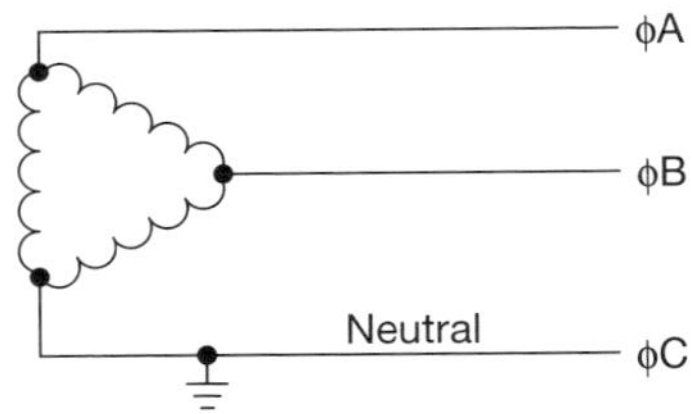

Three-Phase Delta, Three-Wire

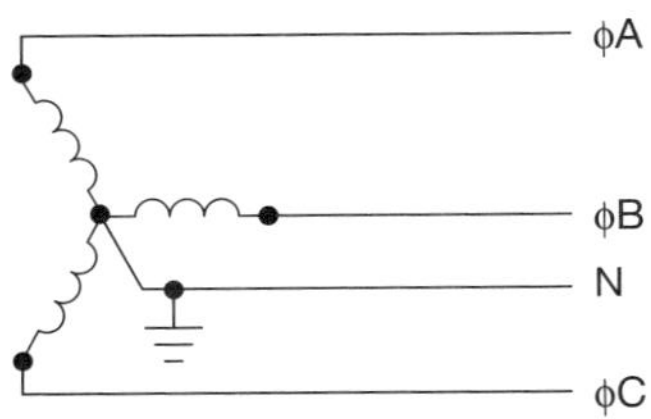

Three-Phase Wye

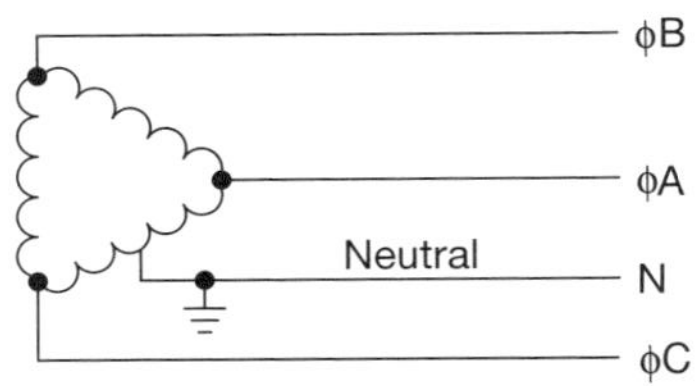

Three-Phase Delta, Four-Wire

GROUNDING DIFFERENT TYPES OF CIRCUITS *(cont.)*

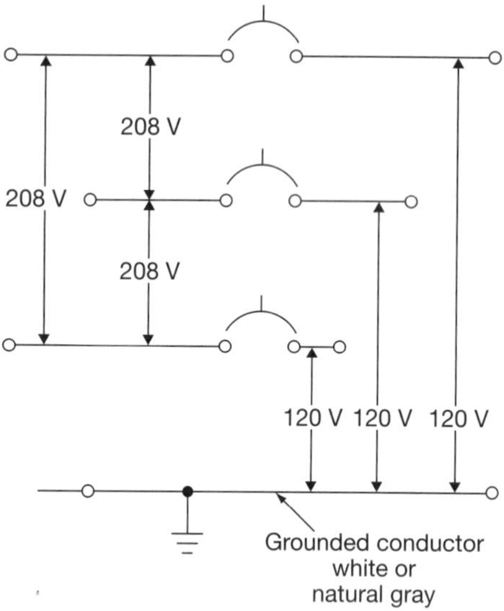

4-Wire System With a Neutral

GROUNDING AN EXISTING CIRCUIT

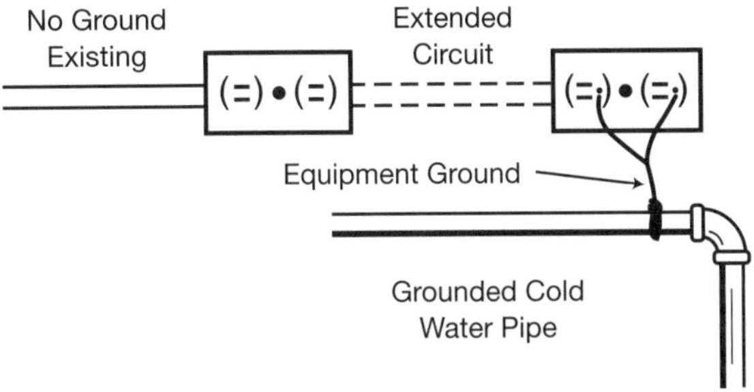

GROUNDING A SCREW-SHELLBASE

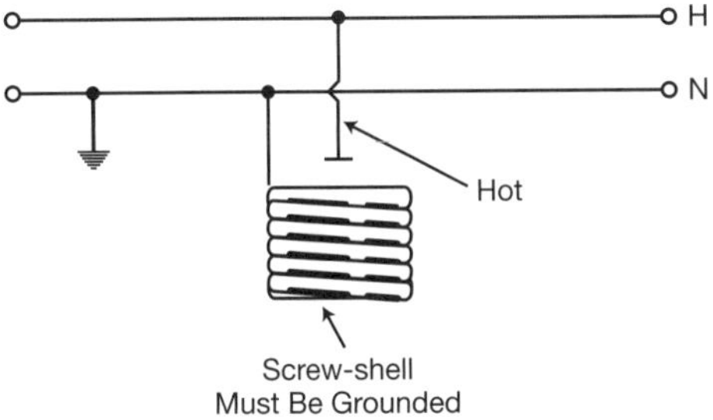

BONDING A TYPICAL SERVICE ENTRANCE

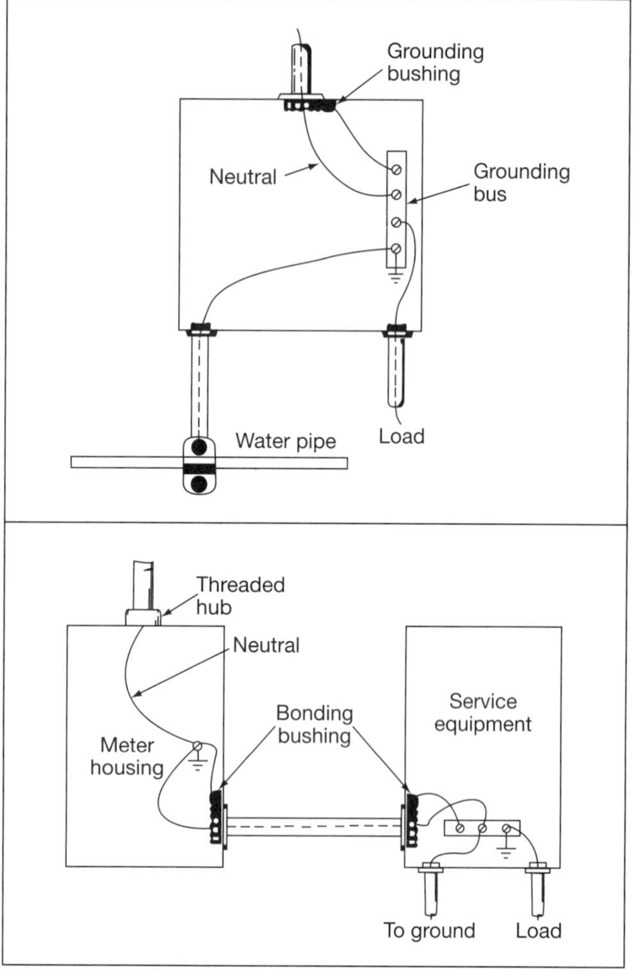

BONDING SERVICE EQUIPMENT METHODS

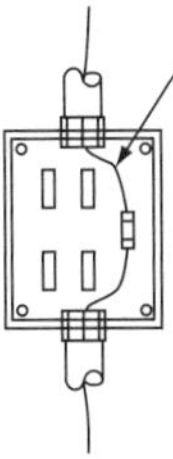

Grounded Service Conductor

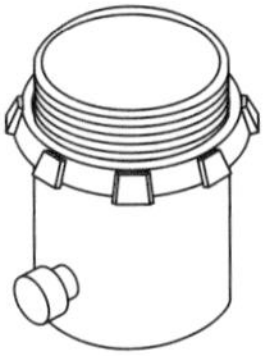

Internal Threadless Connections

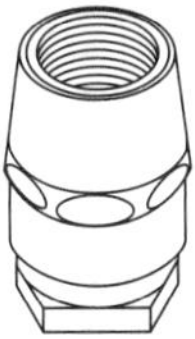

Internal Threaded Connections

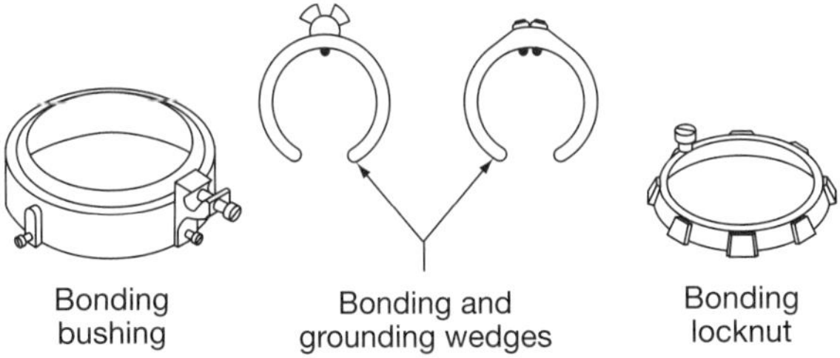

Other Devices

BONDING JUMPERS FOR GROUNDING EQUIPMENT CONDUCTOR RACEWAYS

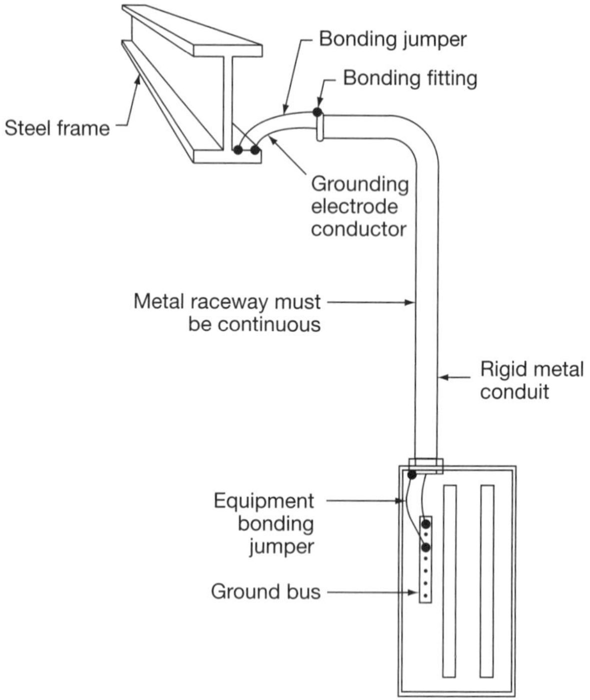

EQUIPMENT BONDING JUMPERS

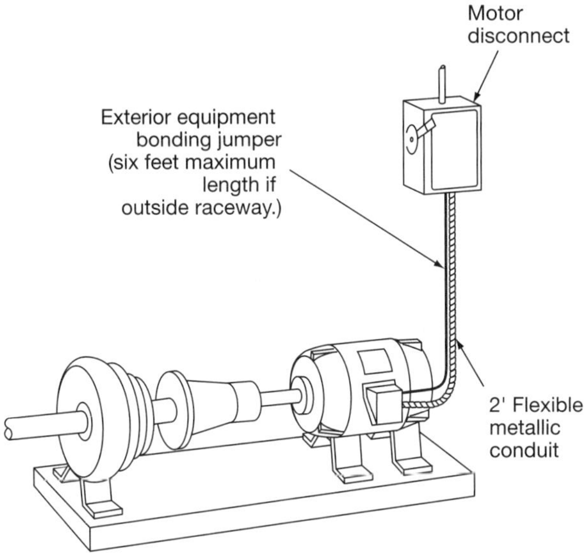

EQUIPMENT GROUNDING CONDUCTORS

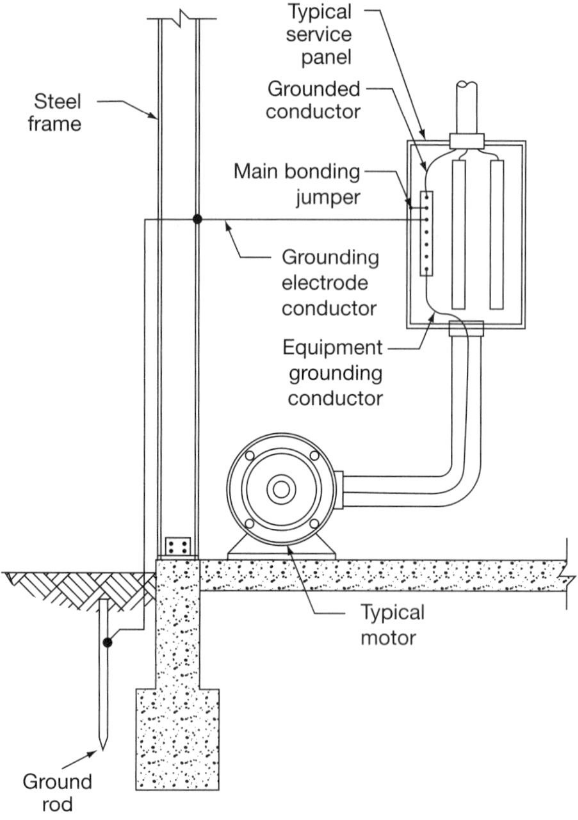

MINIMUM SIZE CONDUCTORS FOR GROUNDING RACEWAY AND EQUIPMENT		
Setting of automatic overcurrent devices in circuits ahead of equipment, conduit, etc., are not to exceed the ampacity ratings below.	Conductor Size (AWG or kcmil)	
	Copper	Aluminum or copper-clad aluminum
15	14 AWG	12 AWG
20	12	10
30	10	8
40	10	8
60	10	8
100	8	6
200	6	4
300	4	2
400	3	1
500	2	1/0
600	1	2/0
800	1/0	3/0
1000	2/0	4/0
1200	3/0	250 kcmil
1600	4/0	350
2000	250 kcmil	400
2500	350	600
3000	400	600
4000	500	800
5000	700	1200
6000	800	1200

BASIC GROUNDING CONNECTIONS

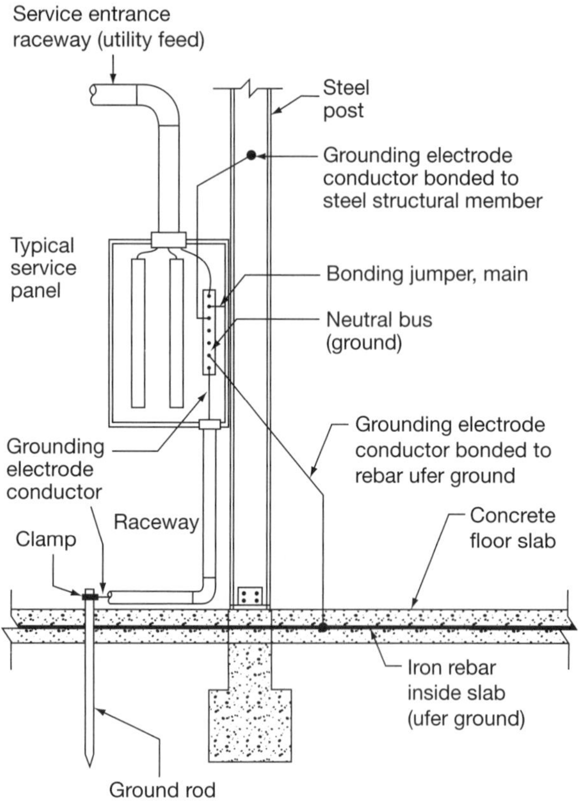

GROUNDING ELECTRODE CONDUCTORS FOR AC SYSTEMS

Service-Entrance Conductor or Equivalent Area for Parallel Conductors		Grounding Electrode Conductor Size and Type	
Copper	**Aluminum or Copper-Clad Aluminum**	**Copper**	**Aluminum or Copper-Clad Aluminum**
2 OR SMALLER 1 OR 1/0 2/0 OR 3/0 AWG	1/0 OR SMALLER 2/0 OR 3/0 AWG 4/0 OR 250 kcmil	8 6 4	6 4 2
OVER 3/0 THRU 350 kcmil	OVER 250 THRU 500 kcmil	2	1/0
OVER 350 kcmil THRU 600 kcmil	OVER 500 kcmil THRU 900 kcmil	1/0	3/0
OVER 600 kcmil THRU 1100 kcmil	OVER 900 kcmil THRU 1750 kcmil	2/0	4/0
OVER 1100 kcmil	OVER 1750 kcmil	3/0	250 kcmil

A) The table above applies to the derived conductors of separately derived AC systems.

B) When multiple sets of service conductors are utilized, the equivalent size of the largest service-entrance conductor shall be determined by the largest sum of the areas of the corresponding conductors of each set.

C) If there are no service-entrance conductors, the grounding electrode conductor size shall be determined by the equivalent size of the largest service-entrance conductor required for the load to be served.

D) Refer to the NEC® installation restrictions concerning aluminum and copper-clad aluminum conductors.

TYPES OF GROUNDING METHODS PER THE NEC®

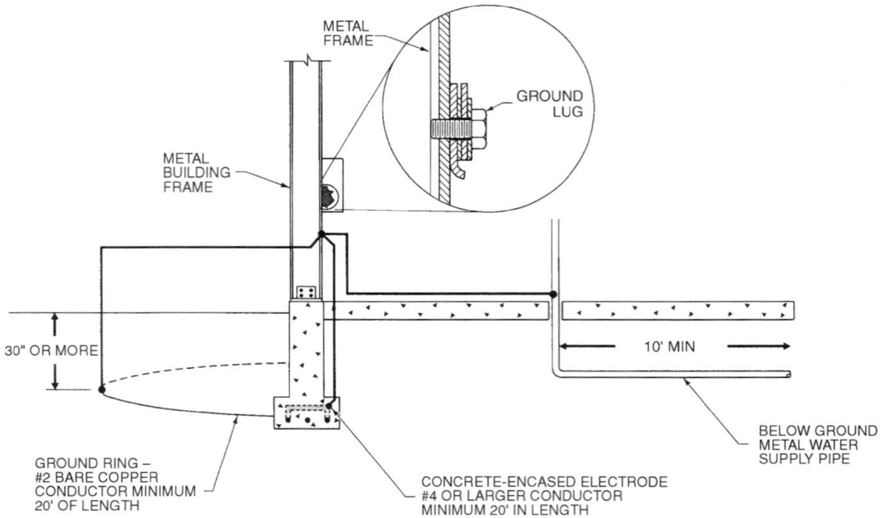

GROUNDING A BASIC NONMETALLIC UNDERGROUND SERVICE RACEWAY

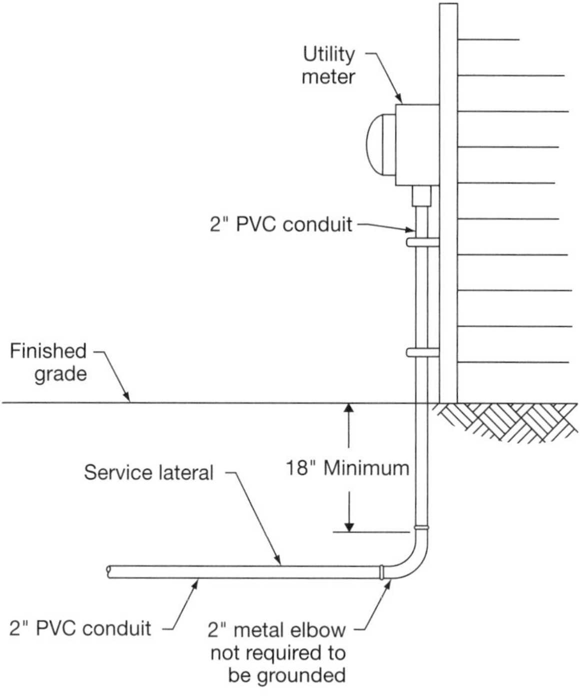

Metal elbows installed in nonmetallic raceways are not required to be grounded as long as they are buried to a minimum of 18" below the finished grade.

GROUNDING A BASIC OVERHEAD ELECTRICAL SERVICE

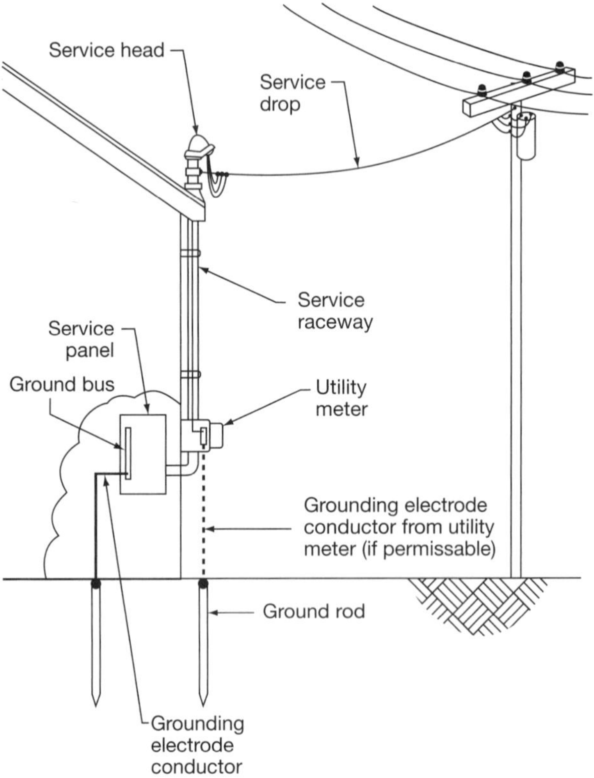

GENERAL REQUIREMENTS FOR GROUNDING AND BONDING SERVICES

A) All grounded AC services for premises shall have a grounding electrode conductor connected to the grounded service conductor.

B) For minimum size of service neutrals use the largest of:
- The maximum unbalanced load
- The size of the grounding electrode conductor.

C) Size the grounding electrode conductor based on the largest service entrance conductor or reference the chart on page 5-15.

D) Size the equipment bonding jumper based on the size of the grounding electrode conductor on the supply side of the service.

E) Run a neutral to each service disconnect and bond the neutral bus to the enclosure.

F) Bond all metal service conduits to enclosures by:
- Using bonding jumpers across knockouts
- Using bonding locknuts for cut holes and outer rings
- The conductor size to ground rods should be 6 AWG.

G) Types of grounding electrodes are:

- Minimum 10' of metal water service pipe in soil
- Grounded structural steel in a building
- Re-bar or steel rod encased in concrete
- Buried metal plates

Note: Two grounding electrodes are required.

H) When bonding to interior metal building structures use:
- Metal water pipes or other metal piping systems
- Structural steel systems

GROUNDING A TYPICAL RESIDENTIAL WIRING SYSTEM

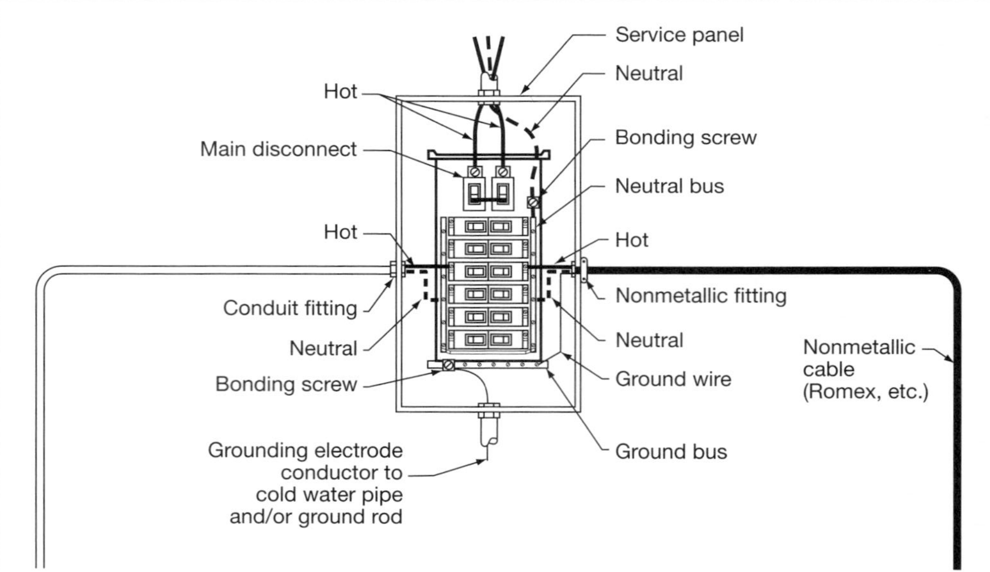

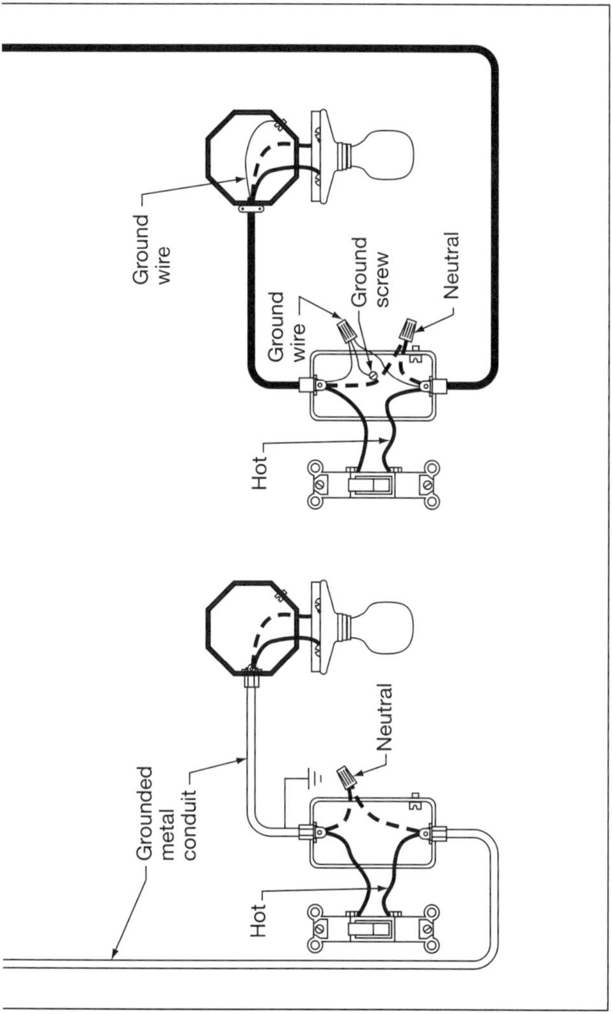
Ground wire
Ground wire
Ground screw
Neutral
Hot
Grounded metal conduit
Neutral
Hot

GROUNDED PRIMARY DISTRIBUTION SYSTEM

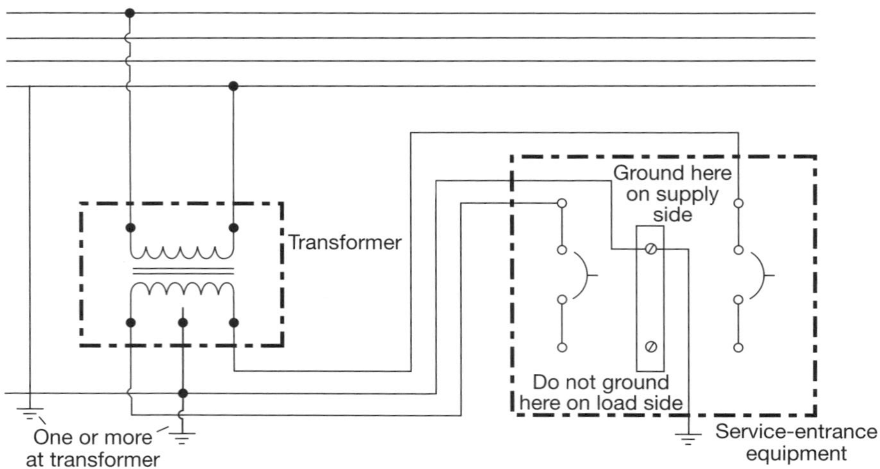

Grounding the supply side will provide a ground if power is ever disconnected.

CHAPTER 6
Motors

DESIGNING MOTOR BRANCH CIRCUITS

For one motor:

1. Determine full-load current of motor(s).
2. Multiply full-load current × 1.25 to determine minimum conductor ampacity.
3. Determine wire size.
4. Determine conduit size.
5. Determine minimum fuse or circuit breaker size.
6. Determine overload rating.

For more than one motor:

1. Perform steps 1 through 6 as shown above for each motor.
2. Add full-load current of all motors, plus 25% of the full-load current of the largest motor to determine minimum conductor ampacity.
3. Determine wire size.
4. Determine conduit size.
5. Add the fuse or circuit breaker size of the largest motor, plus the full-load currents of all other motors to determine the maximum fuse or circuit breaker size for the feeder.

MOTOR BRANCH CIRCUIT REQUIREMENTS

Branch Circuit Conductors:

In general, the conductors that supply a single motor used in a continuous duty application shall have an ampacity of not less than 125% of the motor's highest full-load current rating. For multiple motors add the full-load current of all the motors plus 25% of the highest full-load current of the largest motor.

Motor Disconnecting Means:

An individual disconnecting means shall be provided for each motor controller and must be in sight of both the motor and controller.

Fuses or Circuit Breakers:

Time-delay fuses for most motors are rated at 175% of the full-load amps.

Circuit breakers for most motors are rated at 250% of the full-load amps.

Note: Refer to the chart on overcurrent protective device ratings on page 6-14 for additional information.

MOTOR BRANCH CIRCUIT REQUIREMENTS *(cont.)*

Motor Controllers:

The motor controller shall have the same horsepower rating as the motor at the application voltage. A branch-circuit inverse time circuit breaker and a molded case switch shall be permitted as controllers.

Motor Overloads:

The overload device shall be selected to trip at:

125% — For motors with a service factor of 1.15 or greater or motors with a marked temperature rise of 40°C or less.

115% — For all other motors.

A "thermally protected" motor does not require additional overload protection.

Motor Full-Load Currents (Ampacities):

To size overload protective devices, use the full-load amps rating from the motor nameplate or refer to the charts on motor full-load current ratings located in this chapter on pages 6-5 thru 6-9.

NEMA RATINGS OF AC MOTOR STARTERS IN AMPS PER HP, PHASE AND VOLTAGE

Nema Motor Str. Size	8 Hr. Open Rating in AMPS	Motor Horse Power				
		3ϕ			1ϕ	
		200 V (208)	230 V (240)	460/575 V (480/610)	115 V (120)	230 V (240)
00	9	1½	1½	2	⅓	1
0	18	3	3	5	1	2
1	27	7½	7½	10	2	3
2	45	10	15	25	3	7½
3	90	25	30	50	7½	15
4	135	40	50	100	—	—
5	270	75	100	200	—	—
6	540	150	200	400	—	—
7	810	—	300	600	—	—
8	1,215	—	450	900	—	—
9	2,250	—	800	1600	—	—

FULL-LOAD CURRENTS IN AMPS FOR VARIOUS THREE-PHASE AC MOTORS

	Induction Type Squirrel-Cage and Wound-Rotor							Synchronous Type*			
HP	115 V	200 V	208 V	230 V	460 V	575 V	2300 V	230 V	460 V	575 V	2300 V
½	4.4	2.5	2.4	2.2	1.1	0.9	—	—	—	—	—
¾	6.4	3.7	3.5	3.2	1.6	1.3	—	—	—	—	—
1	8.4	4.8	4.6	4.2	2.1	1.7	—	—	—	—	—
1½	12.0	6.9	6.6	6.0	3.0	2.4	—	—	—	—	—
2	13.6	7.8	7.5	6.8	3.4	2.7	—	—	—	—	—
3	—	11.0	10.6	9.6	4.8	3.9	—	—	—	—	—
5	—	17.5	16.7	15.2	7.6	6.1	—	—	—	—	—
7½	—	25.3	24.2	22.0	11.0	9.0	—	—	—	—	—
10	—	32.2	30.8	28.0	14.0	11.0	—	—	—	—	—
15	—	48.3	46.2	42.0	21.0	17.0	—	—	—	—	—
20	—	62.1	59.4	54.0	27.0	22.0	—	—	—	—	—
25	—	78.2	74.8	68.0	34.0	27.0	—	53.0	26.0	21.0	—
30	—	92.0	88.0	80.0	40.0	32.0	—	63.0	32.0	26.0	—
40	—	120	114	104	52.0	41.0	—	83.0	41.0	33.0	—
50	—	150	143	130	65.0	52.0	—	104	52.0	42.0	—
60	—	177	169	154	77.0	62.0	16.0	123	61.0	49.0	12.0
75	—	221	211	192	96.0	77.0	20.0	155	78.0	62.0	15.0
100	—	285	273	248	124	99.0	26.0	202	101	81.0	20.0
125	—	359	343	312	156	125	31.0	253	126	101	25.0
150	—	414	396	360	180	144	37.0	302	151	121	30.0
200	—	552	528	480	240	192	49.0	400	201	161	40.0
250	—	—	—	—	302	242	60.0	—	—	—	—
300	—	—	—	—	361	289	72.0	—	—	—	—
350	—	—	—	—	414	336	83.0	—	—	—	—
400	—	—	—	—	477	382	95.0	—	—	—	—
450	—	—	—	—	515	412	103	—	—	—	—
500	—	—	—	—	590	472	118	—	—	—	—

Amperages listed are permitted for system voltage ranges of 110–120, 220–240, 440–480, and 550–600 volts.

*For 90% power factor, multiply the amps by 1.10.

*For 80% power factor, multiply the amps by 1.25.

THREE-PHASE AC MOTOR REQUIREMENTS							
Motor HP	Rated Volts	FLC Amps	OCPD Size	Starter Size	Heater Amps	Wire Size	Conduit Size
½	230 460	2.2 1.1	15 15	00 00	2.530 1.265	12 12	¾" ¾"
¾	230 460	3.2 1.6	15 15	00 00	3.680 1.840	12 12	¾" ¾"
1	230 460	4.2 2.1	15 15	00 00	4.830 2.415	12 12	¾" ¾"
1½	230 460	6.0 3.0	15 15	00 00	6.90 3.45	12 12	¾" ¾"
2	230 460	6.8 3.4	15 15	0 00	7.82 3.91	12 12	¾" ¾"
3	230 460	9.6 4.8	20 15	0 0	11.04 5.52	12 12	¾" ¾"
5	230 460	15.2 7.6	30 15	1 0	17.48 8.74	12 12	¾" ¾"
7½	230 460	22.0 11.0	45 20	1 1	25.30 12.65	10 12	¾" ¾"
10	230 460	28.0 14.0	60 30	2 1	32.20 16.10	10 12	¾" ¾"
15	230 460	42.0 21.0	70 40	2 2	48.30 24.15	6 10	1" ¾"

Notes:

1) Overcurrent device may be increased due to starting current and load conditions.
2) Wire size based on 75°C.
3) Overload heater based on motor nameplate.
4) Conduit size based on Rigid Metal Conduit.
5) Wire and conduit size varies per insulation and termination listing.
6) Amperages listed are permitted for system voltage ranges of 220 to 240 and 440 to 480.
7) Calculations apply only to induction type, squirrel-cage, and wound-rotor motors.

THREE-PHASE AC MOTOR REQUIREMENTS *(cont.)*

Motor HP	Rated Volts	FLC Amps	OCPD Size	Starter Size	Heater Amps	Wire Size	Conduit Size
20	230 460	54.0 27.0	100 50	3 2	62.10 31.05	4 10	1" ¾"
25	230 460	68.0 34.0	100 50	3 2	78.20 39.10	4 8	1½" 1"
30	230 460	80.0 40.0	125 70	3 3	92.00 46.00	3 8	1½" 1"
40	230 460	104.0 52.0	175 100	4 3	119.60 59.80	1 6	1½" 1"
50	230 460	130.0 65.0	200 150	4 3	149.50 74.75	2/0 4	2" 1½"
60	230 460	154.0 77.0	250 200	5 4	177.10 88.55	3/0 3	2" 1½"
75	230 460	192.0 96.0	300 200	5 4	220.80 110.40	250 1	2½" 1½"
100	230 460	248.0 124.0	400 200	5 4	285.20 142.60	350 1/0	3" 2"
125	230 460	312.0 156.0	500 250	6 5	358.80 179.40	600 3/0	3½" 2"
150	230 460	360.0 180.0	600 300	6 5	414.00 207.00	700 4/0	4" 2½"

Notes:

1) Overcurrent device may be increased due to starting current and load conditions.
2) Wire size based on 75°C.
3) Overload heater based on motor nameplate.
4) Conduit size based on rigid metal conduit.
5) Wire and conduit size varies per insulation and termination listing.
6) Amperages listed are permitted for system voltage ranges of 220 to 240 and 440 to 480.
7) Calculations apply only to induction type, squirrel-cage and wound-rotor motors.

FULL-LOAD CURRENTS IN AMPS FOR SINGLE-PHASE AC MOTORS				
HP	**115 V**	**200 V**	**208 V**	**230 V**
1/6	4.4	2.5	2.4	2.2
1/4	5.8	3.3	3.2	2.9
1/3	7.2	4.1	4.0	3.6
1/2	9.8	5.6	5.4	4.9
3/4	13.8	7.9	7.6	6.9
1	16.0	9.2	8.8	8.0
1 1/2	20.0	11.5	11.0	10.0
2	24.0	13.8	13.2	12.0
3	34.0	19.6	18.7	17.0
5	56.0	32.2	30.8	28.0
7 1/2	80.0	46.0	44.0	40.0
10	100.0	57.5	55.0	50.0

Note: Amperages listed are permitted for system voltage ranges of 110–120 and 220–240.

SMALL MOTOR GUIDE			
AC, 115 Volt, 60 Hz, Single-Phase			
	FLC in Amps		
Motor HP	**115 V**	**230 V**	**RPM Speed**
1/20	2.50	1.25	1550
1/15	2.80	1.40	1550
1/12	3.20	1.60	850
	4.10	2.05	1550
	2.80	1.40	1725
1/10	4.00	2.00	1050
	3.50	1.75	1550
1/8	3.80	1.90	1140
	2.50	1.25	1725

FULL-LOAD CURRENTS IN AMPS FOR DIRECT CURRENT MOTORS						
HP	90 V	120 V	180 V	240 V	500 V	550 V
¼	4.0	3.1	2.0	1.6	—	—
⅓	5.2	4.1	2.6	2.0	—	—
½	6.8	5.4	3.4	2.7	—	—
¾	9.6	7.6	4.8	3.8	—	—
1	12.2	9.5	6.1	4.7	—	—
1½	—	13.2	8.3	6.6	—	—
2	—	17.0	10.8	8.5	—	—
3	—	25.0	16.0	12.2	—	—
5	—	40.0	27.0	20.0	—	—
7½	—	58.0	—	29.0	13.6	12.2
10	—	76.0	—	38.0	18.0	16.0
15	—	—	—	55.0	27.0	24.0
20	—	—	—	72.0	34.0	31.0
25	—	—	—	89.0	43.0	38.0
30	—	—	—	106.0	51.0	46.0
40	—	—	—	140.0	67.0	61.0
50	—	—	—	173.0	83.0	75.0
60	—	—	—	206.0	99.0	90.0
75	—	—	—	255.0	123.0	111.0
100	—	—	—	341.0	164.0	148.0
125	—	—	—	425.0	205.0	185.0
150	—	—	—	506.0	246.0	222.0
200	—	—	—	675.0	330.0	294.0
Note: Amperages listed are for motors running at base speed.						

DIRECT CURRENT MOTOR REQUIREMENTS						
Motor HP	Full-Load Current in Amps		Minimum Conduit Size in Inches		Minimum Wire Size AWG/Kcmil	
	115 V	230 V	115 V	230 V	115 V	230 V
1	8.4	4.2	½	½	14	14
1½	12.5	6.3	½	½	12	14
2	16.1	8.3	¾	½	10	14
3	23	12.3	¾	½	8	12
5	40	19.8	1	¾	6	10
7½	58	28.7	1¼	1	3	6
10	75	38	1½	1	1	6
15	112	56	2	1¼	2/0	4
20	140	74	2	1½	3/0	1
25	184	92	2½	2	300	1/0
30	220	110	3	2	400	2/0
40	292	146	3½	2½	700	4/0
50	360	180	4	2½	1000	300
60	—	215	—	3	—	400
75	—	268	—	3½	—	600
100	—	355	—	4	—	1000

DC MOTOR PERFORMANCE CHARACTERISTICS

Performance Characteristics	Voltage 10% below Rated Voltage		Voltage 10% above Rated Voltage	
	Shunt	Compound	Shunt	Compound
Starting Torque	−15%	−15%	+15%	+15%
Speed	−5%	−6%	+5%	+6%
Current	+12%	+12%	−8%	−8%
Field Temperature	Increases	Decreases	Increases	Increases
Armature Temperature	Increases	Increases	Decreases	Decreases
Commutator Temperature	Increases	Increases	Decreases	Decreases

MAXIMUM ACCELERATION TIME

Motor Frame Number	Maximum Acceleration Time (in seconds)
48 and 56	8
143–286	10
324–326	12
364–505	15

HORSEPOWER RATINGS FOR 240 VOLT AC SAFETY SWITCHES

General Duty 240 Volt Applications

Rating in Amps	Fusible				Nonfusible	
	Standard		Maximum		Maximum	
	1ϕ	3ϕ	1ϕ	3ϕ	1ϕ	3ϕ
30	1½	3	3	7½	3	7½
60	3	7½	10	15	10	15
100	7½	15	15	30	15	30
200	15	25	—	60	—	60
400	—	50	—	125	—	125
600	—	75	—	200	—	150

Heavy Duty 240 Volt Applications

Rating in Amps	Fusible				Nonfusible	
	Standard		Maximum		Maximum	
	1ϕ	3ϕ	1ϕ	3ϕ	1ϕ	3ϕ
30	1½	3	3	7½	5	10
60	3	7½	10	15	10	20
100	7½	15	15	30	20	40
200	15	25	15	60	—	60
400	—	50	—	125	—	125
600	—	75	—	200	—	200
800	50	—	50	250	—	250
1200	50	—	50	—	—	250

Notes: Horsepower ratings vary by manufacturer. Check the manual or the disconnect itself for the actual rating.

The maximum horsepower rating applies to time delay fuses and the standard to nontime delay fuses.

HORSEPOWER RATINGS FOR 240 VOLT AC SAFETY SWITCHES *(cont.)*

Heavy Duty 480 Volt Applications

Rating in Amps	Fusible		Nonfusible
	Standard	Maximum	Maximum
	3ϕ	3ϕ	3ϕ
30	5	15	20
60	15	30	50
100	25	60	75
200	50	125	125
400	100	250	250
600	150	400	400
800	200	500	500
1200	200	500	500

Notes: Horsepower ratings vary by manufacturer. Check the manual or the disconnect itself for the actual rating.

The maximum horsepower rating applies to time delay fuses and the standard to nontime delay fuses.

Additional Horsepower Rating Notes:

1) For a safety switch to be horsepower rated, it must be able to interrupt a motor's locked rotor current and capable of handling six times the motor's full-load current rating.
2) The horsepower rating of IEC motor starters must be handled differently in that they do not have the same reserve capacity as NEMA® starters.
3) Derate the horsepower if the motor is to perform jogging or plugging duty.
4) When IEC motors are used and are rated in watts, use the following guidelines for converting to equivalent horsepower.
   720 watts = 1 horsepower or
   1 kilowatt = 1.34 horsepower
5) Remember that these figures will vary based on motor efficiency.

MAXIMUM RATING OR SETTING OF MOTOR BRANCH CIRCUIT OVERCURRENT PROTECTIVE DEVICES IN PERCENT OF FULL-LOAD CURRENT[1]

Type of Motor	Nontime Delay Fuse[2]	Dual Element (Time-Delay) Fuse[2]	Instantaneous Trip Breaker	Inverse Time Breaker[3]
Single-phase	300%	175%	800%	250%
AC polyphase except wound-rotor	300%	175%	800%	250%
Squirrel cage except Design B (energy-efficient)	300%	175%	800%	250%
Design B (energy-efficient)	300%	175%	1100%	250%
Synchronous[4]	300%	175%	800%	250%
Wound-rotor	150%	150%	800%	150%
Direct current (constant voltage)	150%	150%	250%	150%

[1] For certain exceptions to the values specified, refer to the NEC®.

[2] The values in the Nontime Delay Fuse column apply to Time-Delay Class CC fuses.

[3] The values given in the last column also cover the ratings of nonadjustable inverse time types of circuit breakers that may be modified (refer to the NEC® for additional information).

[4] Synchronous motors of the low-torque, low-speed type (450 rpm or lower) that start unloaded, do not require a fuse rating or circuit-breaker setting in excess of 200 percent of full-load current.

OVERLOAD DEVICE PLACEMENT		
Type of Motor	**Motor Supply Circuit Type**	**Number and Placement in Circuit for Overload Devices Such as Trip Coils or Relays**
Single-phase AC or DC	2-wire, single-phase AC or DC, ungrounded	1 in either conductor
Single-phase AC or DC	2-wire, single-phase AC or DC, one conductor grounded	1 in ungrounded conductor
Single-phase AC or DC	3-wire, single-phase AC or DC, grounded neutral conductor	1 in either ungrounded conductor
Single-phase AC	Any three-phase	1 in ungrounded conductor
Two-phase AC	3-wire, two-phase AC, ungrounded	2, one in each phase
Two-phase AC	3-wire, two-phase AC, one conductor grounded	2 in ungrounded conductors
Two-phase AC	4-wire, two-phase AC, grounded or ungrounded	2, one per phase in ungrounded conductors
Two-phase AC	Grounded neutral or 5-wire, two-phase AC, ungrounded	2, one per phase in any ungrounded phase wire
Three-phase AC	Any three-phase	3, one in each phase*

**Exception:* Not required where overload protection is provided by other means.

HORSEPOWER FORMULAS

Current and Voltage Known

$$HP = \frac{E \times I \times E_{FF}}{746}$$

where
HP = horsepower
I = current (amps)
E = voltage (volts)
E_{FF} = efficiency
746 = constant

Speed and Torque Known

$$HP = \frac{RPM \times T}{5252}$$

where
HP = horsepower
RPM = revolutions per minute
T = torque (ft-lb)
5252 = constant

EFFICIENCY FORMULAS

Input and Output Power Known

$$E_{FF} = \frac{P_{out}}{P_{in}}$$

where
E_{FF} = efficiency (%)
P_{out} = output power (watts)
P_{in} = input power (watts)

Horsepower and Power Loss Known

$$E_{FF} = \frac{746 \times HP}{(746 \times HP) + W_l}$$

where
E_{FF} = efficiency (%)
746 = constant
HP = horsepower
W_l = watts lost

VOLTAGE UNBALANCE

$$V_u = \frac{V_d}{V_a} \times 100$$

where
V_u = voltage unbalance (%)
V_d = voltage deviation (volts)
V_a = voltage average (volts)
100 = constant

TEMPERATURE CONVERSIONS

Convert °C to °F

$$°F = (1.8 \times °C) + 32$$

Convert °F to °C

$$°C = \frac{(°F - 32)}{1.8}$$

LOCKED ROTOR CURRENT FORMULAS			
Apparent, 1ϕ	**Apparent, 3ϕ**	**True, 1ϕ**	**True, 3ϕ**
$LRC = \frac{1000 \times HP \times kVA/HP}{V}$ where LRC = locked rotor current (in amps) 1000 = multiplier for kilo HP = horsepower kVA/HP = kilovolt amps per horsepower V = volts	$LRC = \frac{1000 \times HP \times kVA/HP}{V \times \sqrt{3}}$ where LRC = locked rotor current (in amps) 1000 = multiplier for kilo HP = horsepower kVA/HP = kilovolt amps per horsepower V = volts $\sqrt{3} = 1.732$	$LRC = \frac{1000 \times HP \times kVA/HP}{V \times PF \times E_{FF}}$ where LRC = locked rotor current (in amps) 1000 = multiplier for kilo HP = horsepower kVA/HP = kilovolt amps per horsepower V = volts PF = power factor E_{FF} = motor efficiency	$LRC = \frac{1000 \times HP \times kVA/HP}{V \times \sqrt{3} \times PF \times E_{FF}}$ where LRC = locked rotor current (in amps) 1000 = multiplier for kilo HP = horsepower kVA/HP = kilovolt amps per horsepower V = volts $\sqrt{3} = 1.732$ PF = power factor E_{FF} = motor efficiency

GEAR REDUCER FORMULAS

Output Torque	Output Speed	Output Horsepower
$O_T = I_T \times R_R \times R_E$ where O_T = output torque (ft-lb) I_T = input torque (ft-lb) R_R = gear reducer ratio R_E = reducer efficiency (%)	$O_s = \frac{I_S}{R_R} \times R_E$ where O_s = output speed (rpm) I_S = input speed (rpm) R_R = gear reducer ratio R_E = reducer efficiency (%)	$O_{HP} = I_{HP} \times R_E$ where O_{HP} = output horsepower I_{HP} = input horsepower R_E = reducer efficiency (%)

MOTOR TORQUE FORMULAS

Torque	Starting Torque	Nominal Torque Rating
$T = \frac{HP \times 5252}{RPM}$ where T = torque (ft-lb) HP = horsepower 5252 = constant RPM = revolutions per minute	$T = \frac{HP \times 5252 \times E_{FF}}{RPM}$ where T = torque (ft-lb) HP = horsepower 5252 = constant RPM = revolutions per minute E_{FF} = motor efficiency (%)	$T = \frac{HP \times 63{,}025}{RPM}$ where T = nominal torque rating (in-lb) HP = horsepower 63,025 = constant RPM = revolutions per minute

MOTOR TORQUE PER MOTOR SPEED

Motor HP	RPM Speeds of Motors			
	68	100	155	190
1	927	630	407	332
1½	1,390	945	610	498
2	1,854	1,261	813	663
3	2,781	1,891	1,220	995
5	4,634	3,151	2,033	1,659
7½	6,951	4,727	3,050	2,488
10	9,268	6,303	4,066	3,317
15	13,903	9,454	6,099	4,976
20	18,537	12,605	8,132	6,634
25	23,171	15,756	10,165	8,293
30	27,805	18,908	12,198	9,951
40	37,074	25,210	16,265	13,268
50	46,342	31,513	20,331	16,586
60	55,610	37,815	24,397	19,903
70	64,879	44,118	28,463	23,220
80	74,147	50,420	32,529	26,537
90	83,415	56,723	36,595	29,854
100	92,684	63,025	40,661	33,171
125	115,855	78,781	50,827	41,464
150	139,026	94,538	60,992	49,757
175	162,197	110,294	71,157	58,049
200	185,368	126,050	81,323	66,342
225	208,539	141,806	91,488	74,635
250	231,710	157,563	101,653	82,928
275	254,881	173,319	111,819	91,220
300	278,051	189,075	121,984	99,513
350	324,393	220,588	142,315	116,099
400	370,735	252,100	162,645	132,684
450	417,077	283,613	182,976	149,270
500	463,419	315,125	203,306	165,855
550	509,761	346,638	223,637	182,441
600	556,103	378,150	243,968	199,026

The torque values above are measured as inch-pounds of force. To obtain torque in foot-pounds of force, divide these values by 12.

MOTOR TORQUE PER MOTOR SPEED *(cont.)*				
Motor HP	**RPM Speeds of Motors**			
	500	**750**	**850**	**1000**
1	126	84	74	63
1½	189	126	111	95
2	252	168	148	126
3	378	252	222	189
5	630	420	371	315
7½	945	630	556	473
10	1,261	840	741	630
15	1,891	1,261	1,112	945
20	2,521	1,691	1,483	1,261
25	3,151	2,101	1,854	1,576
30	3,782	2,521	2,224	1,891
40	5,042	3,361	2,966	2,521
50	6,303	4,202	3,707	3,151
60	7,563	5,042	4,449	3,782
70	8,824	5,882	5,190	4,412
80	10,084	6,723	5,932	5,042
90	11,345	7,563	6,673	5,672
100	12,605	8,403	7,415	6,303
125	15,756	10,504	9,268	7,878
150	18,908	12,605	11,122	9,454
175	22,059	14,706	12,976	11,029
200	25,210	16,807	14,829	12,605
225	28,361	18,908	16,683	14,181
250	31,513	21,008	18,537	15,756
275	34,664	23,109	20,390	17,332
300	37,815	25,210	22,244	18,908
350	44,118	29,412	25,951	22,059
400	50,420	33,613	29,659	25,210
450	56,723	37,815	33,366	28,361
500	63,025	42,017	37,074	31,513
550	69,328	46,218	40,781	34,664
600	75,630	50,420	44,488	37,815

The torque values above are measured as inch-pounds of force. To obtain torque in foot-pounds of force, divide these values by 12.

MOTOR TORQUE PER MOTOR SPEED *(cont.)*

Motor HP	RPM Speeds of Motors 1050	1550	1725	3450
1	60	41	37	18
1½	90	61	55	27
2	120	81	73	37
3	180	122	110	55
5	300	203	183	91
7½	450	305	274	137
10	600	407	365	183
15	900	610	548	274
20	1,200	813	731	365
25	1,501	1,017	913	457
30	1,801	1,220	1,096	548
40	2,401	1,626	1,461	731
50	3,001	2,033	1,827	913
60	3,601	2,440	2,192	1,096
70	4,202	2,846	2,558	1,279
80	4,802	3,253	2,923	1,461
90	5,402	3,660	3,288	1,644
100	6,002	4,066	3,654	1,827
125	7,503	5,083	4,567	2,284
150	9,004	6,099	5,480	2,740
175	10,504	7,116	6,394	3,197
200	12,005	8,132	7,307	3,654
225	13,505	9,149	8,221	4,110
250	15,006	10,165	9,134	4,567
275	16,507	11,182	10,047	5,024
300	18,007	12,198	10,961	5,480
350	21,008	14,231	12,788	6,394
400	24,010	16,265	14,614	7,307
450	27,011	18,298	16,441	8,221
500	30,012	20,331	18,268	9,134
550	33,013	22,364	20,095	10,047
600	36,014	24,397	21,922	10,961

The torque values above are measured as inch-pounds of force. To obtain torque in foot-pounds of force, divide these values by 12.

TIPS ON SELECTING MOTORS

First step in selecting a motor for a particular drive is to obtain data listed below. Motors generally operate at best power factor and efficiency when fully loaded.

TORQUE: Starting torque needed by load must be less than required starting torque of proposed motor. Motor torque must never fall below driven machine's torque needs in going from standstill to full speed.

Torque requirements of some loads may fluctuate between wide limits. Although average torque may be low, many torque peaks may be well above full-load torque. If load torque impulses are repeated frequently (air compressor), it's best to use a high-slip motor with a flywheel. But if load is generally steady at full load, you can use a more efficient low-slip motor. Only in this case any intermittent load peaks are taken directly by the motor and reflect back into power system. Also breakdown (maximum motor torque) must be higher than load-peak torque.

ENCLOSURES: Atmospheric conditions surrounding motor determine enclosure used. The more enclosed a motor, the more it costs and the hotter it tends to run. Totally-enclosed motors may require a larger frame size for a given hp than open or protected motors.

INSULATION: This, likewise, is determined by surrounding atmosphere and operating temperature. Ambient (room) temperature is generally assumed to be 40°C. Total temperature motor reaches directly influences insulation life. Each 10°C rise in Max. temp halves effective life of *Class A* and *B* insulations.

Motor temperature rise is maximum temperature (over ambient) measured with an external thermometer. "Hot-spot" allowance takes care of temperature difference between external reading and hottest spot within windings. Service-factor allows for continuous overload—15% for general-purpose open, protected or drip-proof motors.

VARIABLE CYCLE: Where load varies according to some regular cycle, it would not be economical to select a motor that matches the peak load.

Instead (for AC induction motors where speed is not varied) calculate hp needed on the *root-means-square* (rms) basis. *Rms* hp is equivalent to the continuous hp that would produce the same heat in motor as cycle operation. Torque-speed relation of motor should still match that of load.

FACTS TO CONSIDER

REQUIREMENTS OF DRIVEN MACHINE:

1. HP needed
2. Torque range
3. Operating cycle-frequency of starts and stops
4. Speed
5. Operating position-horizontal, vertical or tilted
6. Direction of rotation
7. Endplay and thrust
8. Ambient (room) temperature
9. Surrounding conditions-water, gas, corrosion, dust, outdoor, etc.

ELECTRICAL SUPPLY:

1. Voltage of power system
2. Number of phases
3. Frequency
4. Limitations on starting current
5. Effect of demand, energy on power rates

DIRECT CURRENT MOTOR DATA

Chief reason for using DC motors, assuming normal power source is AC, lies in the wide and economical ranges possible of speed control and starting torques. But for constant-speed service, AC motors are generally preferred because they are more rugged and have lower first cost.

STARTING TORQUE: With a shunt motor, torque is proportional to armature current, because field flux remains practically constant for a given setting of the field rheostat. However, the flux of a series field is affected by the current through it. At light loads, flux varies directly with a current, so torque varies as the square of the current. The compound motor (usually cumulative) lies in between the shunt and series motors as to torque.

Upper limit of current input on starting is usually 1.5 to 2 times full-load current to avoid over-heating the commutator, excessive fedder drops or peaking generator. Shunt-motor starting boxes usually allow 125% current at first notch. So motor can develop 125% starting torque. Series motors can develop higher starting torques at same current, since torque increases as current squared. Compound motors develop starting torques higher than shunt motors according to amount of compounding.

SPEED CONTROL: Shunt motor speeds drop only slightly (5% or less) from no load to full load. Decreasing field current raises speed; increasing field reduces speed. But speed is still practically constant for any one field setting. Speed can be controlled by resistance in the armature circuit but regulation is poor.

Series motor speeds decrease much more with increased load, and, conversely, begin to race at low loads, dangerously so if load is completely removed. Speed can be reduced by adding resistance into the armature circuit, increased by shunting the series field with resistance or short-circuiting series turns.

Compound motors have less constant speed than shunt motors and can be controlled by shunt-field rheostat.

APPLICATIONS FOR DIRECT CURRENT AND SINGLE-PHASE MOTORS

Speed Regulation	Speed Control	Starting Torque	Pull-out Torque	Motor Applications
Series				
Varies inversely as the load. Races on light loads and full voltage	Zero to maximum depending on control and load	High. Varies as square of the voltage. Limited by commutation, heating and line capacity	High. Limited by commutation, heating and line capacity	Where high starting torque is required and speed can be regulated. Traction, bridges, hoists, gates, car dumpers, car retarders, etc.
Shunt				
Drops 3 to 5% from no load to full load	Any desired range depending on motor design and type of system	Good. With constant field, varies directly as voltage applied to armature	High. Limited by commutation, heating and line capacity	Where constant or adjustable speed is required and starting conditions are not severe. Fan, blowers, centrifugal pumps, conveyers, wood working machines, metal working machines, elevators
Compound				
Drops 7 to 20% from no load to full load depending on amount of compounding	Any desired range depending on motor design and type of control	Higher than for shunt, depending on amount of compounding	High. Limited by commutation, heating and line capacity	Where high starting torque combined with fairly constant speed is required. Plunger pumps, punch presses, shears, bending rolls, geared elevators, conveyors, hoists

APPLICATIONS FOR DIRECT CURRENT AND SINGLE-PHASE MOTORS *(cont.)*

Speed Regulation	Speed Control	Starting Torque	Pull-out Torque	Motor Applications
CAPACITOR				
Drops 5% for large to 10% for small sizes	None	150 to 350% of full load depending upon design and size	150% for large to 200% for small sizes	Constant speed service for any starting duty, and quiet operation, where polyphase current cannot be used
COMMUTATOR-TYPE				
Drops 5% for large to 10% for small sizes	Repulsion-induction, none. Brush shifting types 4 to 1 at full load	250% for large to 350% for small sizes	150% for large to 250% for small sizes	Constant speed service for any starting duty, where speed control is required and polyphase current cannot be used
SPLIT-PHASE				
Drops about 10% from no load to full load	None	75% for large to 175% for small sizes	150% for large to 200% for small sizes	Constant speed service where starting is easy. Small fans, centrifugal pumps, and light running machines, where polyphase current is not available

SQUIRREL-CAGE MOTOR DATA

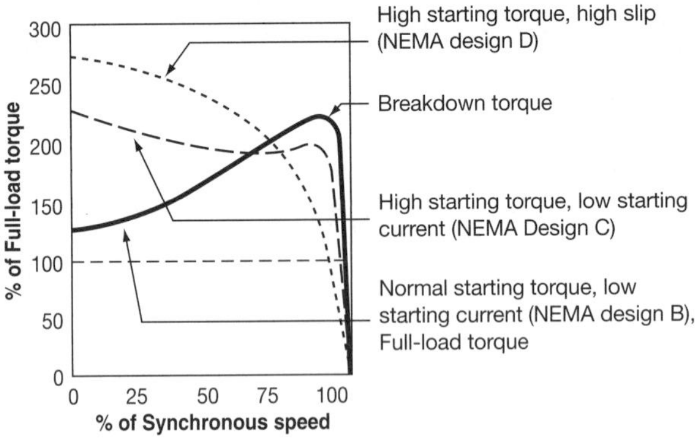

Squirrel-cage induction motors are classed by National Electrical Manufacturers Assn. (NEMA) according to locked-rotor torque, breakdown torque, slip, starting current, etc. Common types are Class B, C and D.

CLASS B is most common type, has normal starting torque, low starting current. Locked-rotor torque (minimum torque at standstill and full voltage) is not less than 100% full-load for 2- and 4-pole motors, 200 hp and less; 40 to 75% for larger 2-pole motors; 50 to 125% for larger 4-pole motors.

CLASS C features high starting torque (locked-rotor over 200%), low starting current. Breakdown torque not less than 190% full-load torque. Slip at full load is between 1½ and 3%.

CLASS D have high slip, high starting torque, low starting current; are used on loads with high intermittent peaks. Driven machine usually has high-inertia flywheel. At no load motor has little slip; when peak load is applied, motor slip increases. Speed reduction lets driven machine absorb energy from flywheel rather than power line.

STARTING Full-voltage, across-the-line starting is used where power supply permits and full-voltage torque and acceleration are not objectionable. Reduction in starting kVA cuts locked-rotor and accelerating torques.

APPLICATIONS FOR SQUIRREL-CAGE MOTORS

Speed Regulation	Speed Control	Starting Torque	Pull-out Torque	Motor Applications
GENERAL-PURPOSE SQUIRREL-CAGE (Class B)				
Drops about 3% for large to 5% for small sizes	None, except multi-speed types designed for 2 to 4 fixed speeds	200% of full load for 2-pole to 105% for 16 pole designs	200% of full load	Constant-speed service where starting torque is not excessive. Fans, blowers, rotary compressors, centrifugal pumps
HIGH TORQUE SQUIRREL-CAGE (Class C)				
Drops about 3% for large to 6% for small sizes	None, except multi-speed types designed for 2 to 4 fixed speeds	250% of full load for high speed to 200% for low speed designs	200% of full load	Constant-speed service where fairly high starting torque is required at infrequent intervals with starting current of about 400% of full load. Reciprocating pumps and compressors, crushers, etc.
HIGH SLIP SQUIRREL-CAGE (Class D)				
Drops about 10 to 15% from no load to full load	None, except multi-speed types designed for 2 to 4 fixed speeds	225 to 300% of full load, depending on speed with rotor resistance	200%. Will usually not stall until loaded to Max. torque, which occurs at standstill	Constant-speed service and high starting torque, if starting is not too frequent, and for taking high peak loads with or without flywheels. Punch presses, shears and elevators, etc.

WOUND-ROTOR MOTOR DATA

The wound-rotor (slip-ring) induction motor's rotor winding connects through slip-rings to an external resistance that is cut in and out by a controller.

RESISTANCE vs TORQUE: Resistance of rotor winding affects torque developed at any speed. A high-resistance rotor gives high starting torque with low starting current. But low slip at full load, good efficiency and moderate rotor heating takes a low-resistance rotor. Left-hand curves show rotor-resistance effect on torque. With all resistance in, R_1, full- load starting torque is developed at less than 150% full-load current. Successively shorting out steps, at standstill, develops about 225% full-load torque at R_4. Cutting out more reduces standstill torque. Motor operates like a squirrel-cage motor when all resistance is shorted out.

SPEED CONTROL: Having resistance left in, decreases speed regulation. Righthand curves for a typical motor show that with only two steps shorted out, motor operates at 65% synchronous speed because motor torque equals load torque at that speed. But if load torque drops to 50%, motor shoots forward to about 65% synchronous speed.

However, slip rings are normally shorted after motor comes up to speed. Or, for short-time peak loads, motor is operated with a step or two of resistance cut in. At light loads, motor runs near synchronous speed. When peak loads come on, speed drops; flywheel effect of motor and load cushions power supply from load peak.

OTHER FEATURES: In addition to high-starting-torque, low-starting-current applications, wound-rotor motors are used (1) for high-inertia loads where high slip losses that would have to be dissipated in the rotor of a squirrel-cage motor, in coming up to speed, can be given off as heat in wound-rotor's external resistance (2) where frequent starting, stopping and speed control are needed (3) for continuous operation at reduced speed. (Example: boiler draft fan — combustion control varies external resistance to adjust speed, damper regulates air flow between step speeds and below 50% speed.)

CONTROLS Used are across-the-line starters with proper protection (fuses, breakers, etc.) and secondary control with 5 to 7 resistance steps.

SYNCHRONOUS MOTOR DATA

Synchronous motors run at a fixed or synchronous speed determined by line frequency and number of poles in machine. (Rpm = 120 x frequency/number or poles.) Speed is kept constant by locking action of an externally excited DC field. Efficiency is 1 to 3% higher than that of same-size-and-speed induction or DC motors. Also synchronous motors can be operated at power factors from 1.0 down to 0.2 leading for plant power-factor correction. Standard ratings are 1.0 and 0.9 leading PF; machines rated down near 0.2 leading are called synchronous condensers.

STARTING: Pure synchronous motors are not self-starting; so in practice they are built with damper or amortisseur windings. With the field coil shorted through discharge resistor, damper winding acts like a squirrel-cage rotor to bring motor practically to synchronous speed; then field is applied and motor pulls into synchronism, providing motor has developed sufficient "pull-in" torque. Once in synchronism, motor keeps constant speed as long as load torque does not exceed maximum or "pull-out" torque; then machine drops out of synchronism. Driven machine is usually started without load. Low-speed motors may be direct connected.

FIELD AND PF: While motors are rated for specific power factors . . . at constant power, increasing DC field current causes power factor to lead, decreasing field tends to make PF lag. But either case increases copper losses.

TYPES: Polyphase synchronous motors in general use are: (1) high-speed motors, 500 rpm up, (a) general-purpose, 500 to 1800 rpm, 200 hp and below, (b) high-speed units over 200 hp including most 2-pole motors (2) low-speed motors below 500 rpm, (3) special high-torque motors.

COSTS: Small high-speed synchronous motors cost more than comparable induction motors, but with low-speed and large high-speed motor, costs favor the synchronous motor. Cost of leading-PF motors increases approximately inversely proportional to the decrease from unity power factor.

APPLICATIONS FOR WOUND-ROTOR AND SYNCHRONOUS MOTORS

Speed Regulation	Speed Control	Starting Torque	Pull-out Torque	Motor Applications
WOUND-ROTOR				
With rotor rings short circuited, drops about 3% for large to 5% for small sizes	Speed can be reduced to 50% by rotor resistance to obtain stable operation. Speed varies inversely as load	Up to 300% depending on external resistance in rotor circuit and how distributed	200% when rotor slip rings are short circuited	Where high starting torque with low starting current or where limited speed control is required. Fans, centrifugal and plunger pumps, compressors, conveyers, hoists, cranes
SYNCHRONOUS				
Constant	None, except special motors designed for 2 fixed speeds	40% for slow to 160% for medium speed 80% PF designs. Special designs develop higher torques	Unity-of motors 170%; 80% PF motors 225%. Special designs up to 300%	For constant-speed service, direct connection to slow speed machines and where power factor correction is required

TYPES OF MOTOR ENCLOSURES

OPEN-TYPE
has full openings in frame and endbells for maximum ventilation, is lowest cost enclosure.

SEMI-PROTECTED
has screens in top openings to keep out falling objects. PROTECTED has screens in bottom too.

DRIP-PROOF
has upper parts covered to keep out drippings falling at angle not over 15° from vertical.

SPLASH-PROOF
is baffled at bottom to keep out particles coming at angle not over 100° from vertical.

TOTALLY-ENCLOSED
can be non-ventilated, separately ventilated, or explosion proof for hazardous atmospheres.

FAN-COOLED
totally-enclosed motor has double covers. Fan, behind vented outer shroud, is run by motor.

TYPES OF INSULATION

CLASS A (cotton, silk, paper or other organics impregnated with insulating varnish) is considered standard for most applications, allows 105° C total temperature.

40°C ambient
40°C rise by thermometer
15°C "hot-spot" allowance
10°C service factor
105°C total temperature

CLASS B (mica, asbestos, fiber-glass, other inorganics) allows 130°C total temperature.

40°C ambient
70°C rise by thermometer
20°C "hot-spot" allowance
130°C total temperature

CLASS H (including silicone family) is for special high-temperature applications.

SPECIAL CLASS A is highly resistant, but not "proof," against severe moisture, dampness; conductive, corrosive or abrasive dusts and vapors.

TROPICAL is for excessive moisture, high ambients, corrosion, fungus, vermin, insects.

MOTOR FRAME LETTERS

Letter	Designation
G	Gasoline pump motor
K	Sump pump motor
M and N	Oil burner motor
S	Standard short shaft for direct connection
T	Standard dimensions established
U	Previously used as frame designation
Y	Special mounting dimensions required
Z	Denotes shaft extension

SHAFT COUPLING SELECTIONS

Coupling Number	Rated Torque (inch-lbs.)	Maximum Shock Torque (inch-lbs.)
10-101-A	16	45
10-102-A	36	100
10-103-A	80	220
10-104-A	132	360
10-105-A	176	480
10-106-A	240	660
10-107-A	325	900
10-108-A	525	1450
10-109-A	875	2450
10-110-A	1250	3500
10-111-A	1800	5040
10-112-A	2200	6160

MOTOR V-BELTS

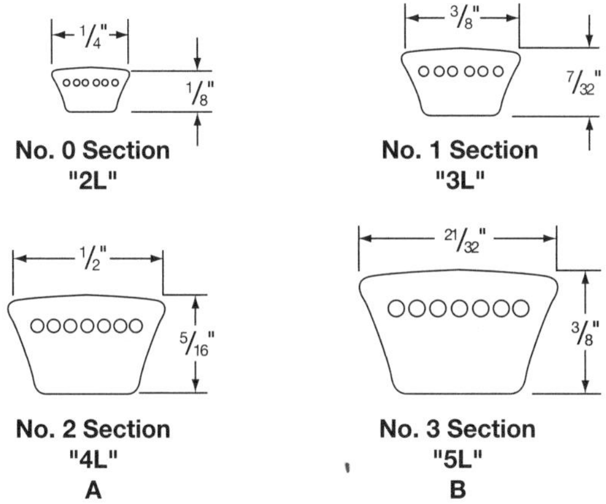

V-BELT PER MOTOR SIZE

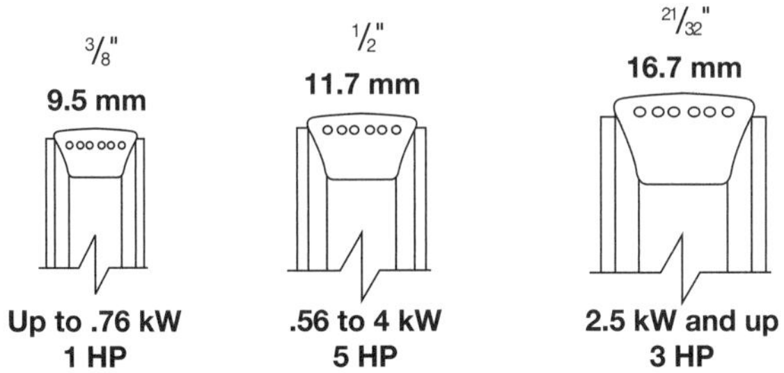

NEMA MOTOR FRAME DIMENSIONS

Frame No.	All Dimension in Inches						
	D	E	F	U	V	M+N	Keyway
42	2⅝	1¾	27/32	⅜	-	4 1/32	-
48	2	2⅛	1⅜	½	-	5⅜	-
56	3½	2 7/16	1½	⅝	-	6⅛	3/16 × 3/32
66	4⅛	2 15/16	2½	¾	-	7⅞	3/16 × 3/32
143T	3½	2¾	2	⅞	2	6½	3/16 × 3/32
145T	3½	2¾	2½	⅞	2	7	3/16 × 3/32
182	4½	3¾	2¼	⅞	2	7¼	3/16 × 3/32
182T	4½	3¾	2¼	1⅛	2½	7¾	¼ × ⅛
184	4½	3¾	2¾	⅞	2	7¾	3/16 × 3/32
184T	4½	3¾	2¾	1⅛	2½	8¼	¼ × ⅛
213	5¼	4¼	2¾	1⅛	2¾	9¼	¼ × ⅛
213T	5¼	4¼	2¾	1⅜	3⅛	9⅝	5/16 × 5/32
215	5¼	4¼	3½	1⅛	2¾	10	¼ × ⅛
215T	5¼	4¼	3½	1⅜	3⅛	10⅜	5/16 × 5/32
254T	6¼	5	4⅛	1⅝	3¾	12⅜	⅜ × 3/16
254U	6¼	5	4⅛	1⅜	3½	12⅛	5/16 × 5/32
256T	6¼	5	5	1⅝	3¾	13¼	⅜ × 3/16
256U	6¼	5	5	1⅜	3½	13	5/16 × 5/32
284T	7	5½	4¾	1⅞	4⅜	14⅛	½ × ¼
284TS	7	5½	4¾	1⅝	3	12¾	⅜ × 3/16
284U	7	5½	4¾	1⅝	4⅝	14⅜	⅜ × 3/16
286T	7	5½	5½	1⅞	4⅜	14⅞	½ × ¼
286U	7	5½	5½	1⅝	4⅝	15⅛	⅜ × 3/16
324T	8	6¼	5¼	2⅛	5	15¾	½ × ¼
324U	8	6¼	5¼	1⅞	5⅜	16⅛	½ × ¼
326T	8	6¼	6	2⅛	5	16½	½ × ¼
326TS	8	6¼	6	1⅞	3½	15	½ × ¼
326U	8	6¼	6	1⅞	5⅜	16⅞	½ × ¼
364T	9	7	5⅝	2⅜	5⅝	17⅜	⅝ × 5/16
364U	9	7	5⅝	2⅛	6⅛	17⅞	½ × ¼
365T	9	7	6⅛	2⅜	5⅝	17⅞	⅝ × 5/16
365U	9	7	6⅛	2⅛	6⅛	18⅜	½ × ¼
404T	10	8	6⅛	2⅞	7	20	¾ × ⅜
404U	10	8	6⅛	2⅜	6⅞	19⅞	⅝ × 5/16
405T	10	8	6⅞	2⅞	7	20¾	¾ × ⅜
405U	10	8	6⅞	2⅜	6⅞	20⅝	⅝ × 5/16
444T	11	9	7¼	3⅜	8¼	23¼	⅞ × 7/16
444U	11	9	7¼	2⅞	8⅜	23⅜	¾ × ⅜
445T	11	9	8¼	3⅜	8¼	24¼	⅞ × 7/16
445U	11	9	8¼	2⅞	8⅜	24⅜	¾ × ⅜

Standards established by National Electrical Manufacturers Association.

FRONTAL VIEW OF TYPICAL MOTOR

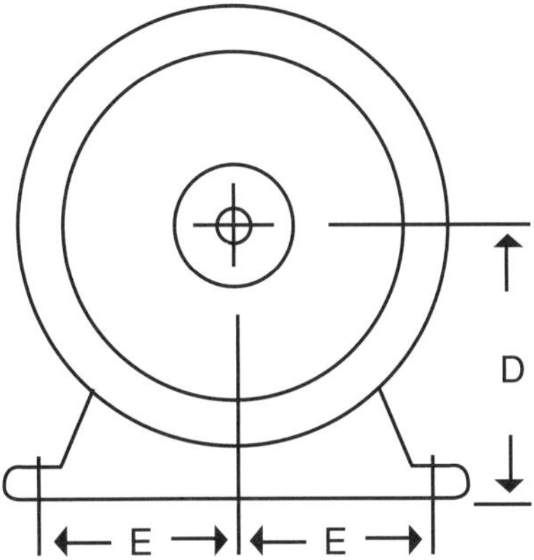

REFERENCE PAGE 6–34 FOR DIMENSIONS

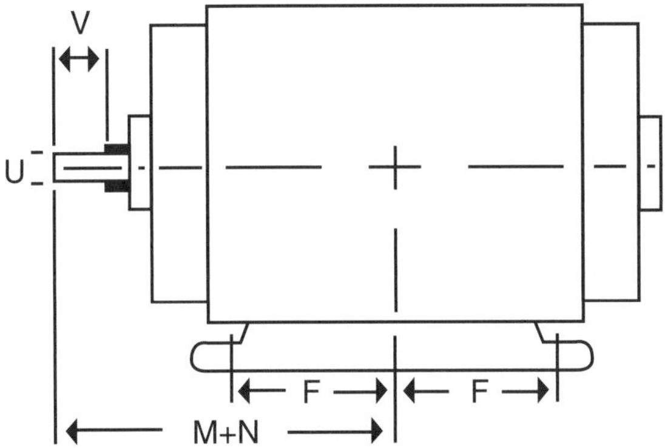

SIDE VIEW OF TYPICAL MOTOR

ROTATION AND TERMINAL MARKINGS FOR DIRECT CURRENT MOTORS

Terminal markings identify connections from outside circuits. Facing the commutator end (the end opposite the drive), the direction of the shaft rotation is counterclockwise. The standard rotation of a direct current generator is clockwise.

A_1 and A_2 are the armature leads.
S_1 and S_2 are the series-field leads.
F_1 and F_2 are the shunt-field leads.

FOR SERIES WOUND MOTORS:
To change rotation, reverse either armature leads or series leads. Do not reverse both.

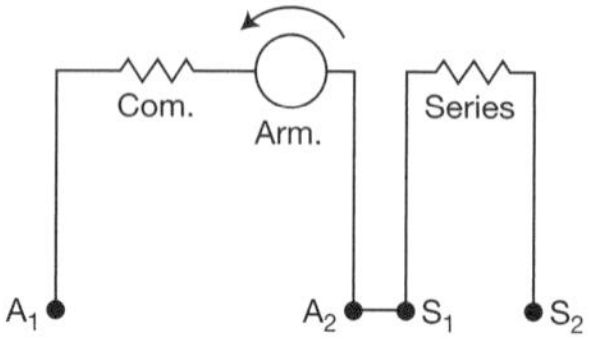

FOR SHUNT WOUND MOTORS:
To change rotation, reverse either armature leads or shunt leads. Do not reverse both.

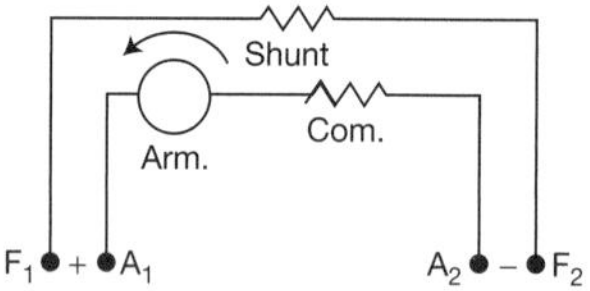

FOR COMPOUND WOUND MOTORS:
To change rotation, reverse either armature leads or both the series and shunt leads. Do not reverse all three.

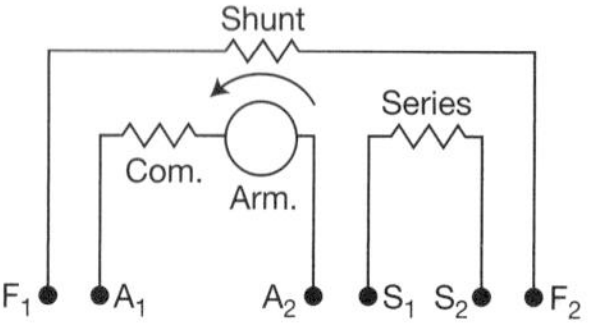

ROTATION AND TERMINAL MARKINGS FOR SINGLE-PHASE MOTORS

There are two main groups of single-phase motors, each of which is made up of various types outlined below.

A) Split-phase:
- Capacitor-Start
- Resistance-Start
- Split-Capacitor

B) Commutator:
- Repulsion
- Repulsion-Start
- Series

The colors of the motor terminals are as follows:

T_1 – Blue	T_3 – Orange	T_5 – Black
T_2 – White	T_4 – Yellow	T_8 – Red

Note: To change the speed of a split-phase motor, the number of poles must be changed, but this is rarely done in the field.

SPLIT-PHASE, SQUIRREL-CAGE MOTORS

Dual Voltage

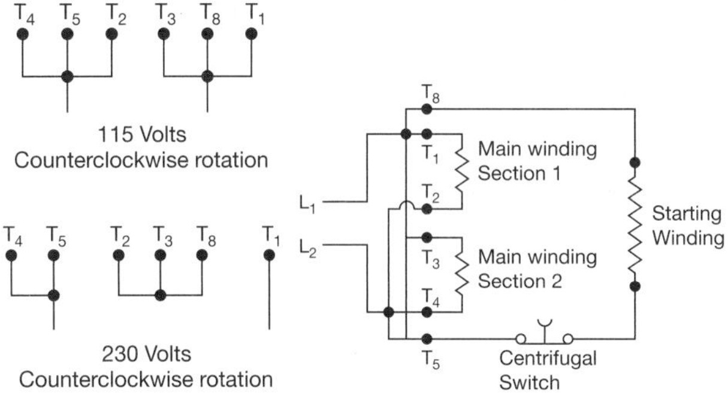

To reverse rotation, interchange T_5 and T_8

Resistance-Start

This motor has a resistance connected in series with the starting winding.

The centrifugal switch opens after reaching 75% of normal operating speed.

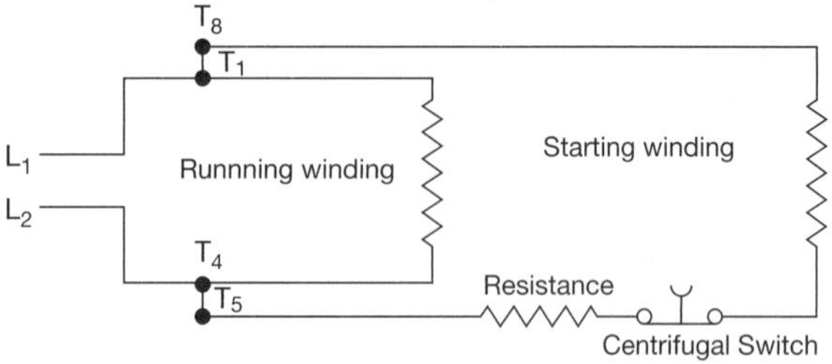

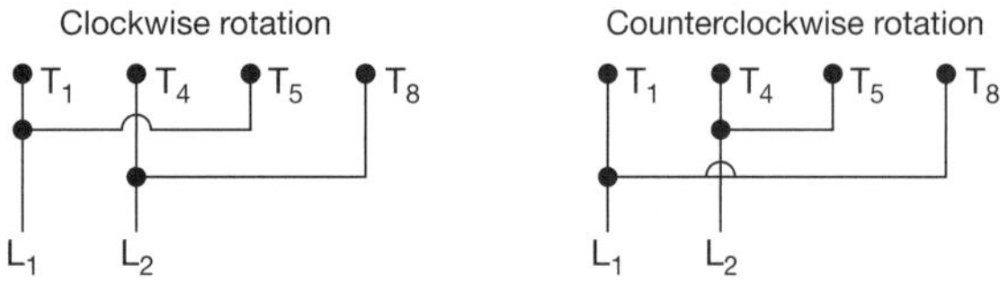

Capacitor-Start

This motor is employed where a high starting torque is required.

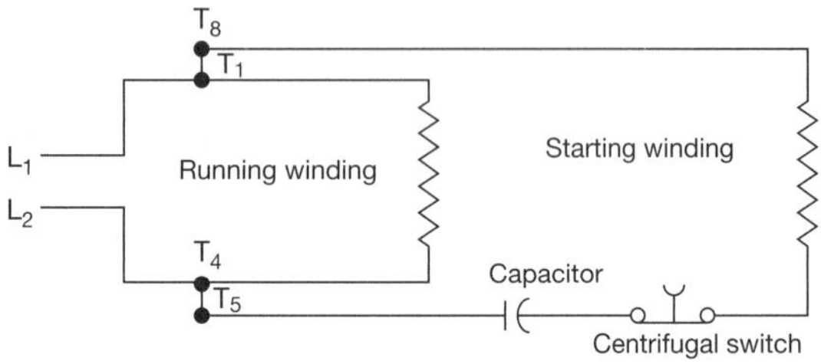

SPLIT-PHASE MOTOR ROTATION

Single Voltage

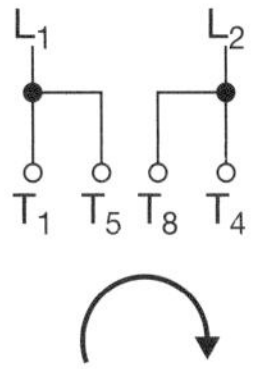

Clockwise Rotation

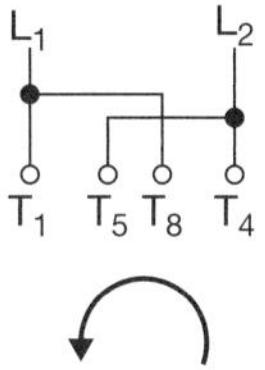

Counterclockwise Rotation

Dual Voltage

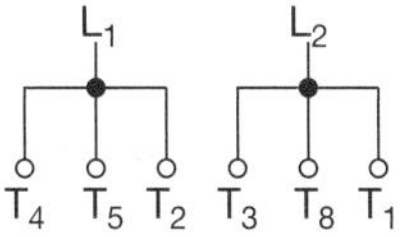

Low voltage

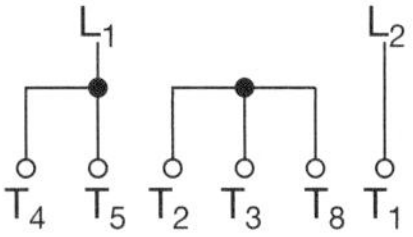

High voltage

Clockwise Rotation

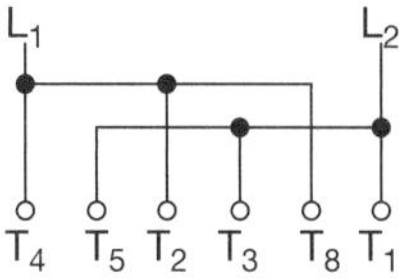

Low voltage

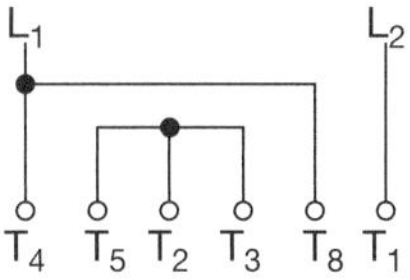

High voltage

Counterclockwise Rotation

DELTA-WOUND MOTOR CONNECTIONS 240/480 V

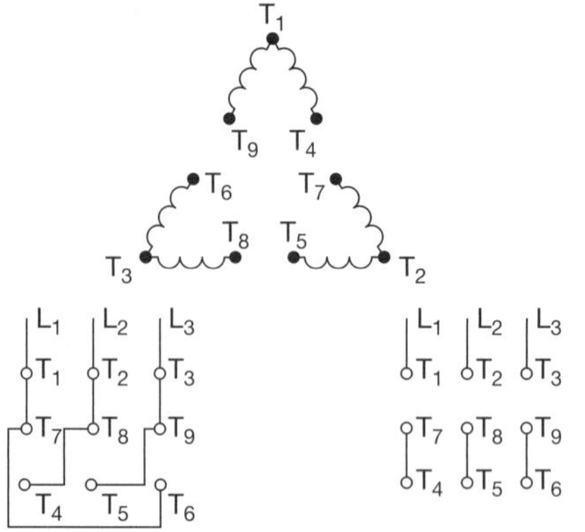

Low-Voltage Connection | **High-Voltage Connection**

WYE-WOUND MOTOR CONNECTIONS 240/480 V

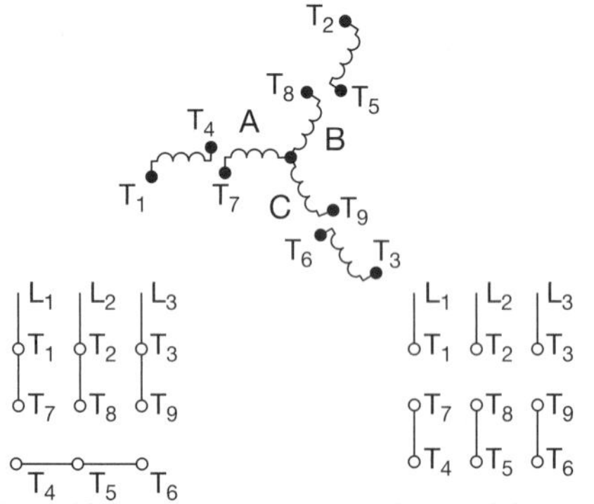

Low-Voltage Connection | **High-Voltage Connection**

DUAL-VOLTAGE, 3ϕ, WYE-CONNECTED MOTORS

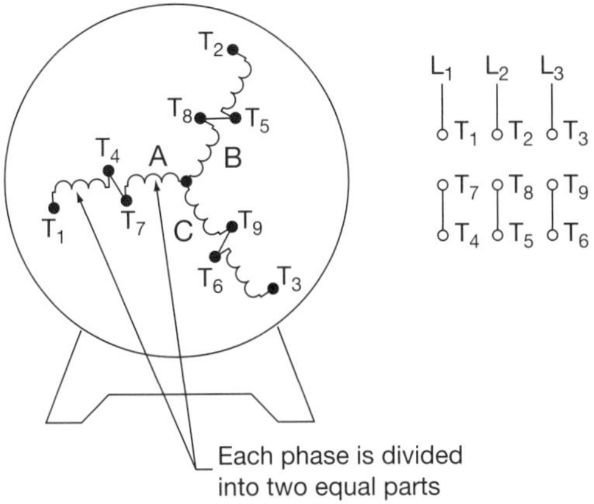

High-Voltage (series)

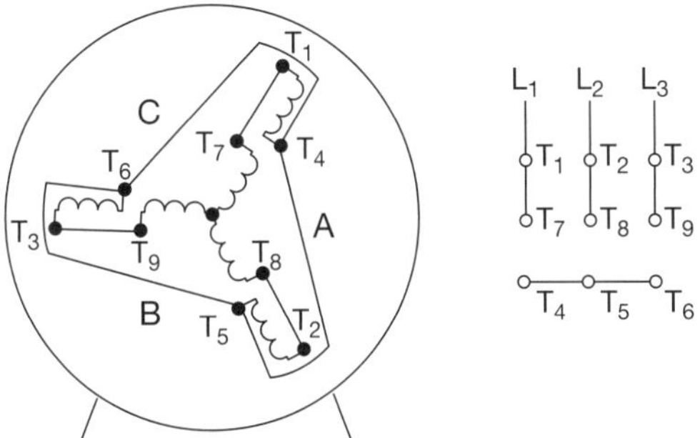

Low-Voltage (parallel)

DUAL-VOLTAGE, 3ϕ, DELTA-CONNECTED MOTORS

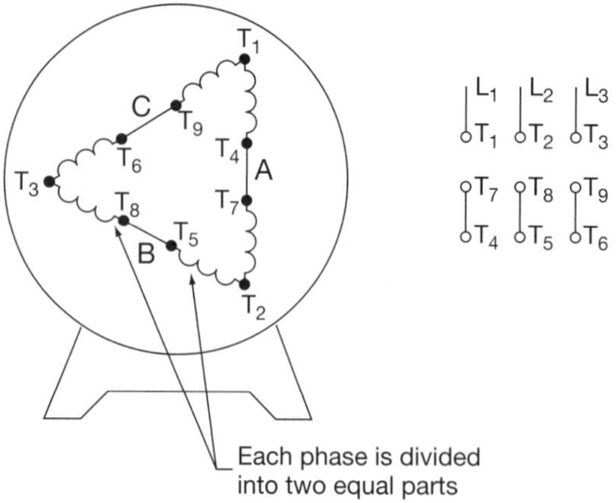

High-Voltage (series)

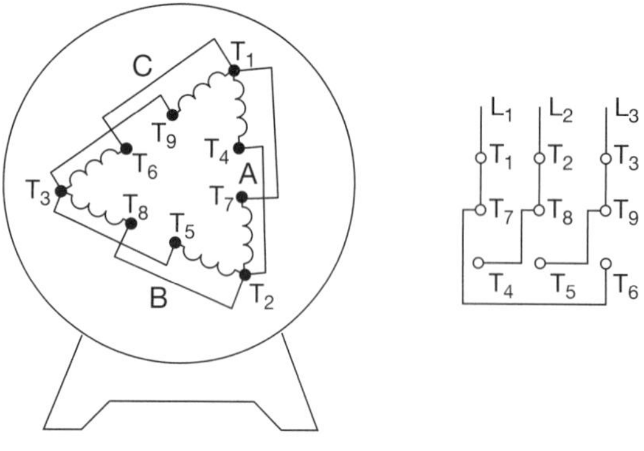

Low-Voltage (parallel)

STAR-CONNECTED, POLYPHASE MOTOR

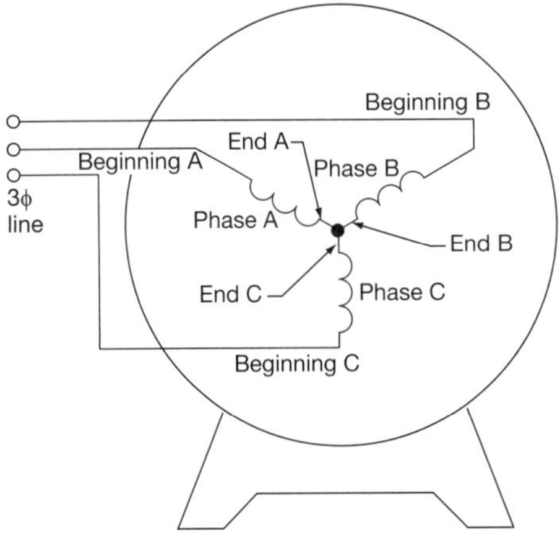

REVERSING THREE-PHASE MOTORS

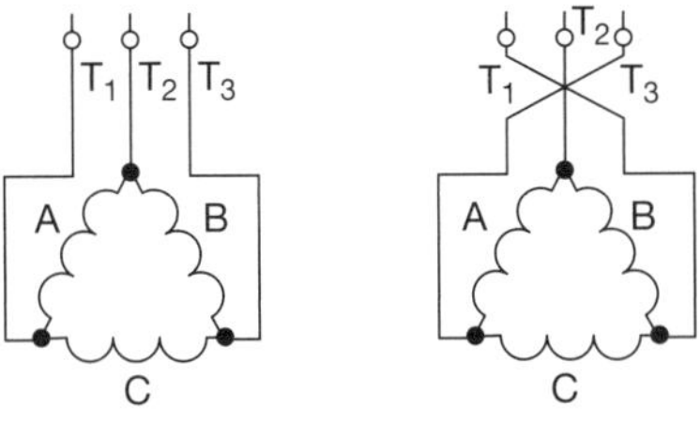

Forward **Reverse**

Reversing direction is accomplished by interchanging any two of the three power lines. Although any two may be interchanged, the industry standard is to interchange L_1 and L_3. This is true for all 3ϕ motors including three, six, and nine lead wye- and delta-connected motors.

MOTOR CONTROL CIRCUIT SCHEMATICS

Magnetic starter with one stop-start station and a pilot lamp which burns to indicate that the motor is running.

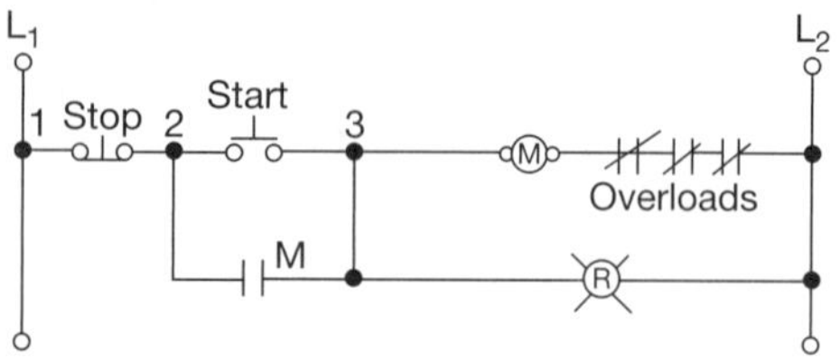

Magnetic starter with three stop-start stations.

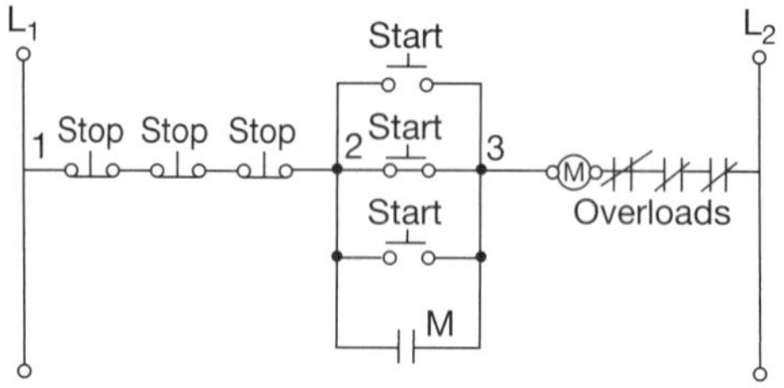

WIRING DIAGRAM OF A SINGLE-PHASE MOTOR CONNECTED TO A THREE-PHASE MOTOR STARTER

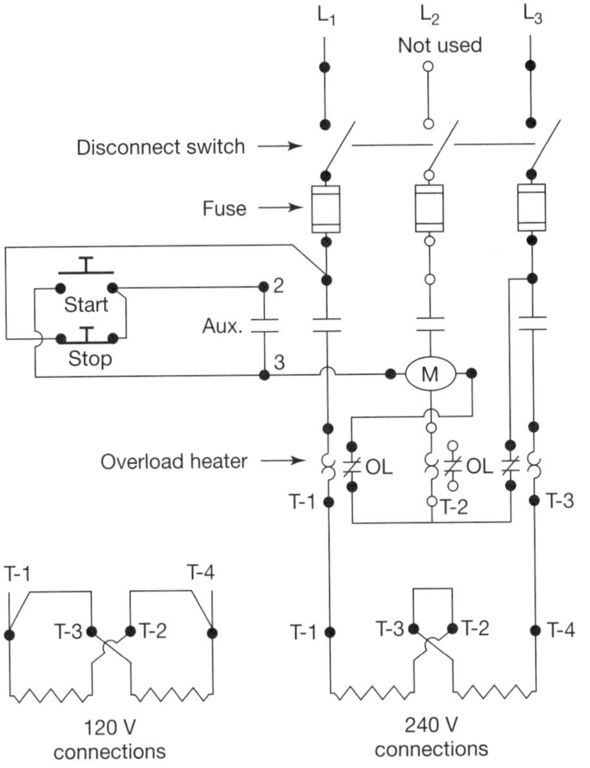

WIRING DIAGRAM OF HAND OFF MOTOR CONTROL

Note: The controls and the motor operate on the same voltage.

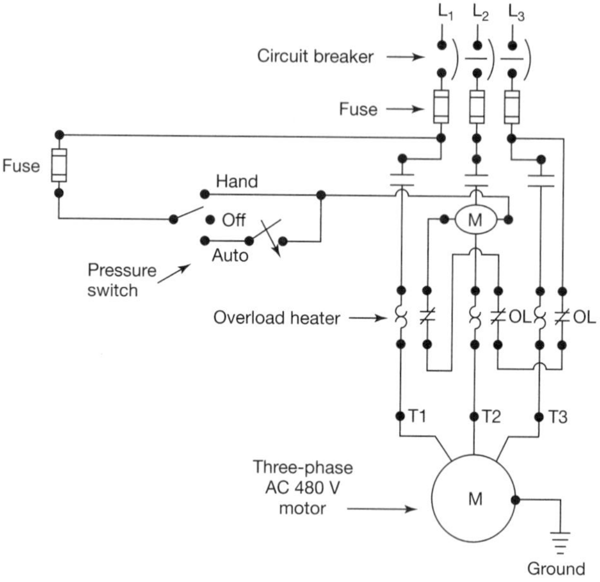

WIRING DIAGRAM OF JOGGING WITH CONTROL RELAY

The jogging circuit allows the motor starter to be energized only when the jog button is depressed.

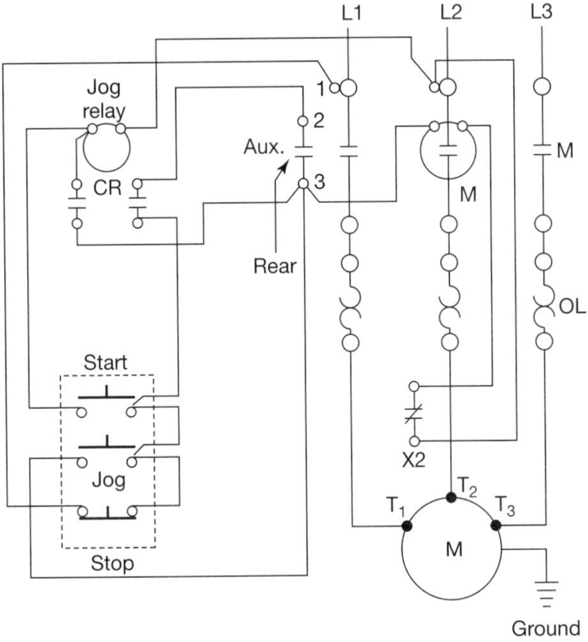

CONTACTOR AND MOTOR STARTER TROUBLESHOOTING GUIDE

Problem	Possible Cause	Corrective Action
Humming noise	Magnet pole faces misaligned	Realign. Replace magnet assembly if realignment is not possible.
	Too low voltage at coil	Measure voltage at coil. Check voltage rating of coil. Correct any voltage that is 10% less than coil rating.
	Pole face obstructed by foreign object, dirt, or rust	Remove any foreign object and clean as necessary. Never file pole faces.
Loud buzz noise	Shading coil broken	Replace coil assembly.
Controller fails to drop out	Voltage to coil not being removed	Measure voltage at coil. Trace voltage from coil to supply looking for shorted switch or contact if voltage is present.
	Worn or rusted parts causing binding	Clean rusted parts. Replace worn parts.
	Contact poles sticking	Checking for burning or sticky substance on contacts. Replace burned contacts. Clean dirty contacts.
	Mechanical interlock binding	Check to ensure interlocking mechanism is free to move when power is OFF. Replace faulty interlock.
Controller fails to pull in	No coil voltage	Measure voltage at coil terminals. Trace voltage loss from coil to supply voltage if voltage is not present.
	Too low voltage	Measure voltage at coil terminals. Correct voltage level if voltage is less than 10% of rated coil voltage. Check for a voltage drop as large loads are energized.
	Coil open	Measure voltage at coil. Remove coil if voltage is present and correct but coil does not pull in. Measure coil resistance for open circuit. Replace if open.

CONTACTOR AND MOTOR STARTER TROUBLESHOOTING GUIDE *(cont.)*

Problem	Possible Cause	Corrective Action
Controller fails to pull in	Coil shorted	Shorted coil may show signs of burning. The fuse or breakers should trip if coil is shorted. Disconnect one side of coil and reset if tripped. Remove coil and check resistance for short if protection device does not trip. Replace shorted coil. Replace any coil that is burned.
	Mechanical obstruction	Remove any obstructions.
Contacts badly burned or welded	Too high inrush current	Measure inrush current. Check load for problem if higher-than-rated load current. Change to larger controller if load current is correct but excessive for controller.
	Too fast load cycling	Change to larger controller if load cycled ON and OFF repeatedly.
	Too large overcurrent protection device	Size overcurrent protection to load and controller.
	Short circuit	Check fuses or breakers. Clear any short circuit.
	Insufficient contact pressure	Check to ensure contacts are making good connection.
Nuisance tripping	Incorrect overload size	Check size of overload against rated load current. Size up if permissible per NEC®.
	Lack of temperature compensation	Correct setting of overload if controller and load are at different ambient temperatures.
	Loose connections	Check for loose terminal connection.

DIRECT CURRENT MOTOR TROUBLESHOOTING GUIDE		
Problem	**Possible Cause**	**Corrective Action**
Motor will not start	Blown fuse or open CB	Test the OCPD. If voltage is present at the input, but not the output of the OCPD, the fuse is blown or the CB is open. Check the rating of the OCPD. It should be at least 125% of the motor's FLC.
	Motor overload on starter tripped	Allow overloads to cool. Reset overloads. If reset overloads do not start motor, test the starter.
	No brush contact	Check brushes. Replace, if worn.
	Open control circuit between incoming power and motor	Check for cleanliness, tightness, and breaks. Use a voltmeter to test the circuit starting with the incoming power and moving to the motor terminals. Voltage generally stops at the problem area.
Fuse, CB, or overloads retrip after service	Excessive load	If the motor is loaded to excess or is jammed, the circuit OCPD will open. Disconnect the load from the motor. If the motor now runs properly, check the load. If the motor does not run and the fuse or CB opens, the problem is with the motor or control circuit. Remove the motor from the control circuit and connect it directly to the power source. If the motor runs properly, the problem is in the control circuit. Check the control circuit. If the motor opens the fuse or CB again, the problem is in the motor. Replace or service the motor.
	Motor shaft does not turn	Disconnect the motor from the load. If the motor shaft still does not turn, the bearings are frozen. Replace or service the motor.
Brushes chip or break	Brush material is too weak or the wrong type for motor's duty rating	Replace with better grade or type of brush. Consult manufacturer if problem continues.
	Brush face is overheating and losing brush bonding material	Check for an overload on the motor. Reduce the load as required. Adjust brush holder arms.

DIRECT CURRENT MOTOR TROUBLESHOOTING GUIDE *(cont.)*

Problem	Possible Cause	Corrective Action
Brushes chip or break	Brush holder is too far from commutator	Too much space between the brush holder and the surface of the commutator allows the brush end to chip or break. Set correct space between brush holder and commutator.
	Brush tension is incorrect	Adjust brush tension so the brush rides freely on the commutator.
Brushes spark	Worn brushes	Replace worn brushes. Service the motor if rapid brush wear, excessive sparking, chipping, breaking, or chattering is present.
	Commutator is concentric	Grind commutator and undercut mica. Replace commutator if necessary.
	Excessive vibration	Balance armature. Check brushes. They should be riding freely.
Rapid brush wear	Wrong brush material, type, or grade	Replace with brushes recommended by manufacturer.
	Incorrect brush tension	Adjust brush tension so the brush rides freely on the commutator.
Motor overheats	Improper ventilation	Clean all ventilation openings, Vacuum or blow dirt out of motor with low-pressure, dry, compressed air.
	Motor is overloaded	Check the load for binding. Check shaft straightness. Measure motor current under operating conditions. If the current is above the listed current rating, remove the motor. Remeasure the current under no-load conditions. if the current is excessive under load but not when unloaded, check the load. If the motor draws excessive current when disconnected, replace or service the motor.

THREE-PHASE MOTOR TROUBLESHOOTING GUIDE		
Problem	**Possible Cause**	**Corrective Action**
Motor will not start	Wrong motor connections	Most 3ϕ motors are dual-voltage. Check for proper motor connections.
	Blown fuse or open CB	Test the OCPD. If voltage is present at the input, but not the output of the OCPD, the fuse is blown or the CB is open. Check the rating of the OCPD. It should be at least 125% of the motor's FLC.
	Motor overload on starter tripped	Allow overloads to cool. Reset overloads. If reset overloads do not start the motor, test the starter.
	Low or no voltage applied to motor	Check the voltage at the motor terminals. The voltage must be present and within 10% of the motor nameplate voltage. If voltage is present at the motor but the motor is not operating, remove the motor from the load the motor is driving. Reapply power to the motor. If the motor runs, the problem is with the load. If the motor does not run, the problem is with the motor. Replace or service the motor.
	Open control circuit between incoming power and motor	Check for cleanliness, tightness, and breaks. Use a voltmeter to test the circuit starting with the incoming power and moving to the motor terminals. Voltage generally stops at the problem area.
Fuse, CB, or overloads retrip after service	Power not applied to all three lines	Measure voltage at each power line. Correct any power supply problems.
	Blown fuse or open CB	Test the OCPD. If voltage is present at the input, but not the output of the OCPD, the fuse is blown or the CB is open. Check the rating of the OCPD. It should be at least 125% of the motor's FLC.
	Motor overload on starter tripped	Allow overloads to cool. Reset overloads. If reset overloads do not start the motor, test the starter.

THREE-PHASE MOTOR TROUBLESHOOTING GUIDE *(cont.)*

Problem	Possible Cause	Corrective Action
Fuse, CB, or over-loads retrip after service	Low or no voltage applied to motor	Check the voltage at the motor terminals. The voltage must be present and within 10% of the motor nameplate voltage. If voltage is present at the motor but the motor is not operating, remove the motor from the load the motor is driving. Reapply power to the motor. If the motor runs, the problem is with the load. If the motor does not run, the problem is with the motor. Replace or service the motor.
	Open control circuit between incoming power and motor	Check for cleanliness, tightness, and breaks. Use a voltmeter to test the circuit starting with the incoming power and moving to the motor terminals. Voltage generally stops at the problem area.
	Motor shaft does not turn	Disconnect the motor from the load. If the motor shaft still does not turn, the bearings are frozen. Replace or service the motor.
Motor overheats	Motor is single phasing	Check each of the 3ϕ power lines for correct voltage.
	Improper ventilation	Clean all ventilation openings. Vacuum or blow dirt out of motor with low-pressure, dry, compressed air.
	Motor is overloaded	Check the load for binding. Check shaft straightness. Measure motor current under operating conditions. If the current is above the listed current rating, remove the motor. Remeasure the current under no-load conditions. If the current is excessive under load but not when unloaded, check the load. If the motor draws excessive current when disconnected, replace or service the motor.

SPLIT-PHASE MOTOR TROUBLESHOOTING GUIDE

Problem	Possible Cause	Corrective Action
Motor will not start	Thermal cutout switch is open	Reset the thermal switch. Caution: Resetting the thermal switch may automatically start the motor.
	Blown fuse or open CB	Test the OCPD, if voltage is present at the input, but not the output of the OCPD, the fuse is blown or the CB is open. Check the rating of the OCPD. It should be at least 125% of the motor's FLC.
	Motor overload on starter tripped	Allow overloads to cool. Reset overloads. If reset overloads do not start the motor, test the starter.
	Low or no voltage applied to motor	Check the voltage at the motor terminals. The voltage must be present and within 10% of the motor nameplate voltage. If voltage is present at the motor but the motor is not operating, remove the motor from the load the motor is driving. Reapply power to the motor. If the motor runs, the problem is with the load. If the motor does not run, the problem is with the motor. Replace or service the motor.
	Open control circuit between incoming power and motor	Check for cleanliness, tightness, and breaks. Use a voltmeter to test the circuit starting with the incoming power and moving to the motor terminals. Voltage generally stops at the problem area.
	Starting winding not receiving power	Check the centrifugal switch to make sure it connects the starting winding when the motor is OFF.
Fuse, CB, or overloads retrip after service	Blown fuse or open CB	Test the OCPD. If voltage is present at the input, but not the output of the OCPD, the fuse is blown or the CB is open. Check the rating of the OCPD. It should be at least 125% of the motor's FLC.
	Motor overload on starter tripped	Allow overloads to cool. Reset overloads. If reset overloads do not start the motor, test the starter.

SPLIT-PHASE MOTOR TROUBLESHOOTING GUIDE *(cont.)*

Problem	Possible Cause	Corrective Action
Fuse, CB, or overloads retrip after service	Low or no voltage applied to motor	Check the voltage at the motor terminals. The voltage must be present and within 10% of the motor nameplate voltage. If voltage is present at the motor but the motor is not operating, remove the motor from the load the motor is driving. Reapply power to the motor. If the motor runs, the problem is with the load. If the motor does not run, the problem is with the motor. Replace or service the motor.
	Open control circuit between incoming power and motor	Check for cleanliness, tightness, and breaks. Use a voltmeter to test the circuit starting with the incoming power and moving to the motor terminals. Voltage generally stops at the problem area.
	Motor shaft does not turn	Disconnect the motor from the load. If the motor shaft still does not turn, the bearings are frozen. Replace or service the motor.
Motor produces electric shock	Broken or disconnected ground strap	Connect or replace ground strap. Test for proper ground.
	Hot power lead at motor connecting terminals is touching motor frame	Disconnect the motor. Open the motor terminal box and check for poor connections, damaged insulation, or leads touching the frame. Service and test motor for ground.
	Motor winding shorted to frame	Remove, service, and test motor.
Motor overheats	Starting windings are not being removed from circuit as motor accelerates	When the motor is turned OFF, a distinct click should be heard as the centrifugal switch closes.
	Improper ventilation	Clean all ventilation openings. Vacuum or blow dirt out of motor with low-pressure, dry, compressed air.

SPLIT-PHASE MOTOR TROUBLESHOOTING GUIDE *(cont.)*

Problem	Possible Cause	Corrective Action
Motor overheats	Motor is overloaded	Check the load for binding. Check shaft straightness. Measure motor current under operating conditions. If current is above the listed current rating, remove the motor. Remeasure the current under no-load conditions. If the current is excessive under load but not when unloaded, check the load. If the motor draws excessive current when disconnected, replace or service the motor.
	Dry or worn bearings	Dry or worn bearings cause noise. The bearings may be dry due to dirty oil, oil not reaching the shaft, or motor overheating. Oil the bearings as recommended. If noise remains, replace the bearings or the motor.
	Dirty bearings	Clean or replace bearings.
Excessive noise	Excessive end-play	Check and play by trying to move the motor shaft in and out. Add end-play washers as required.
	Unbalanced motor or load	An unbalanced motor or load causes vibration, which causes noise. Realign the motor and load. Check for excessive end play or loose parts. If the shaft is bent, replace the rotor or motor.
	Dry or worn bearings	Dry or worn bearings cause noise. The bearings may be dry due to dirty oil, oil not reaching the shaft, or motor overheating. Oil the bearings as recommended. If noise remains, replace the bearings or the motor.
	Excessive grease	Ball bearings that have excessive grease may cause the bearings to overheat. Overheated bearings cause noise. Remove any excess grease.

CHAPTER 7
Transformers

SIZING TRANSFORMER CIRCUITS

1. Determine current in primary windings.

   Example: 75kVA, 3-phase, 480/208V, wye-connected transformer. Full-load current = 75,000VA ÷ 480V = 156 Amperes ÷ 1.732 = 90.2 amperes per phase.
2. Refer to Table 450.3(B) of the NEC to determine size of overcurrent protective device.

   Example: 90.2 × 125% = 112.75 amperes.
3. Choose next-higher standard overcurrent device.

   Example: 125 amperes, per Section 240.6(A).
4. Size primary conductors per overcurrent device.

   Example: #2 THHN conductors, per Table 310.15(B)(16).
5. Determine if secondary overcurrent protection is required.

   Example: Secondary protection required, per Section 240.21(C)(1).
6. Determine secondary current.

   Example: 75,000VA ÷ 208V = 360.6 Amperes ÷ 1.732 = 208 amperes per phase.
7. Refer to Table 450.3(B) of the NEC to determine size of overcurrent protective device.

   Example: 208 × 125% = 260 amperes.
8. Choose next-higher standard overcurrent device.

   Example: 300 amperes, per Section 240.6(A).
9. Size secondary conductors per overcurrent device.

   Example: 300kcmil THHN conductors, per Table 310.15(B)(16).
10. Size main bonding jumper at secondary disconnect, per 250.102(C) and 250.66.

    Example: #2 copper, per Table 250.66.
11. Size grounding electrode conductor.

    Example: #2 copper, per Table 250.66.
12. Choose grounding electrode per Section 250.30(A)(7) and job conditions.

THREE-PHASE CONNECTIONS

Wye Connection

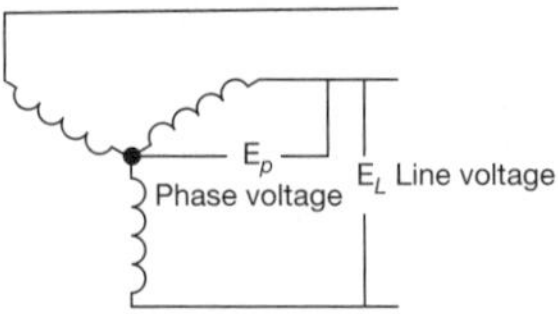

The line current and phase current are the same:

$I_{LINE} = I_{PHASE}$

The line voltage is higher than the phase voltage by a factor of the square root of 3:

$E_{LINE} = E_{PHASE} \times \sqrt{3}$ or $E_{PHASE} = \frac{E_{LINE}}{\sqrt{3}}$

Delta Connection

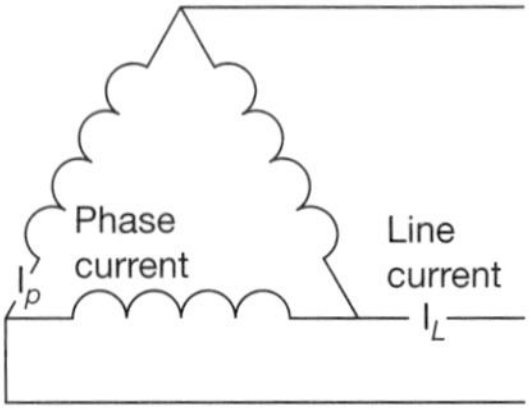

The line voltage and phase voltage are the same:

$E_{LINE} = E_{PHASE}$

The line current is higher than the phase current by a factor of the square root of$\sqrt{3}$:

$I_{LINE} = I_{PHASE} \times \sqrt{3}$ or $I_{PHASE} = \frac{I_{LINE}}{\sqrt{3}}$

Open Delta Connection

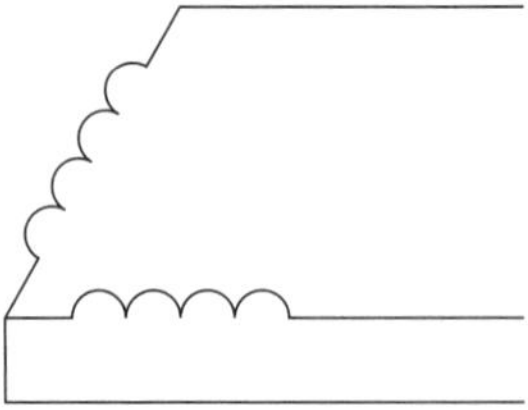

Open-delta connections provide 86.6% of the sum of the power rating of two transformers. Example: Two transformers are rated at 60 kVA each. The total power rating for this connection would be as follows:

$60 + 60 = 120$ kVA

$120 \text{ kVA} \times 0.866 = 103.92$ kVA

Voltage and current are computed in the same manner as for a closed-delta connection.

TRANSFORMER TURNS RATIO

A transformer increases or decreases voltage by inducing electrical energy from one coil to another through magnetic lines of force. The turns ratio is the ratio between the voltage and the number of turns on the primary and secondary windings of a transformer.

Thus

$$\frac{N_p}{N_s} = \frac{E_p}{E_s}$$

where

N_s = number of turns in secondary

E_p = primary voltage

E_s = secondary voltage

Transformer with 2:1
Voltage Ratio

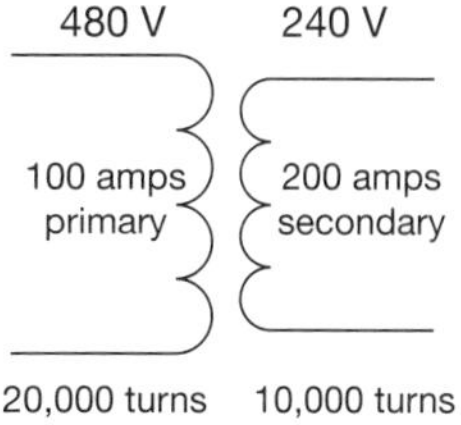

Primary Voltage = 2 times secondary voltage

Secondary Current = 2 times primary current

Primary Winding Turns = 2 Times Secondary Winding Turns

FEEDER TAP AND TRANSFORMER TAP INSTALLATION GUIDELINES

Feeder conductors shall be permitted to be tapped without overcurrent protection as follows (see NEC 240.21 for all details):

Up to a 10 Ft. Feeder Tap

- Feeder may be tapped with a smaller conductor.
- Ampacity of tap conductor not less than the combined load, the rating of device supplied or the rating of the overcurrent protection at end of tap conductors.

Up to a 25 Ft. Feeder Tap

- Feeder may be tapped with a smaller conductor.
- Ampacity of tap conductor not less than ⅓ of the rating of the overcurrent protection of feeder.

Conductors shall be permitted to be connected to a transformer secondary without overcurrent protection as follows:

Secondary Conductors Up to 10 Ft.

- Ampacity of conductors not less than the combined load, the rating of device supplied or the rating of the overcurrent protection at end of secondary conductors.

Secondary Conductors Up to 25 Ft.

- If the conductors on the primary are protected at their ampacity, the conductors on the secondary may be run up to 25 ft. before overcurrent protection is needed.

Outdoor Transformer Tap

- There is no limit to the conductor length.
- Single main must be provided where conductors enter a building.

TRANSFORMER FEED TAP NOT OVER 25 FEET LONG

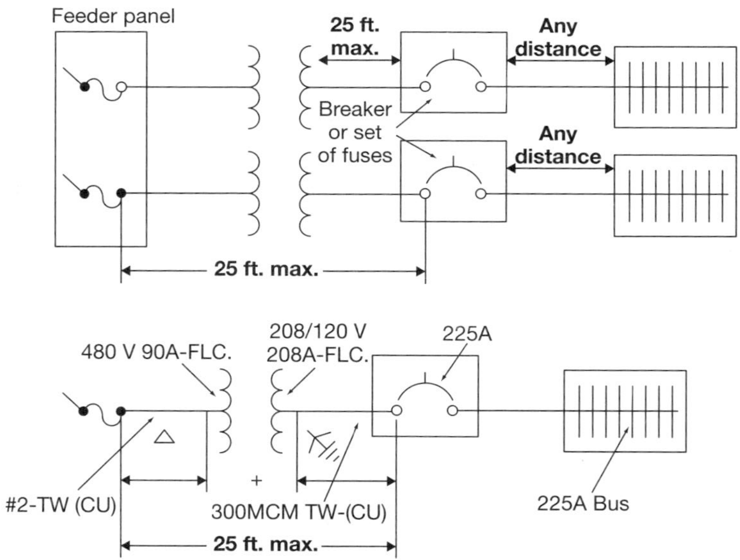

See NEC 240.21 for all details

SIZING SINGLE-PHASE TRANSFORMERS

1. Determine the total voltage required by the loads.
2. Determine the kVA capacity required by the loads.
3. Check frequency of supply voltage and the loads.
4. Check supply voltage with rating of primary side.

To calculate kVA capacity of a 1ϕ transformer

$$kVA_{CAP} = E \times \frac{I}{1000}$$

Where: kVA_{CAP} = transformer capacity (in kVA)
E = voltage (in Volts)
I = current (in Amps)

The transformer must have a kVA capacity 10% greater than that required by the loads.

kVA Rating	1ϕ FULL-LOAD CURRENTS (AMPS)				
	120 V	**208 V**	**240 V**	**480 V**	**2400 V**
1	8.33	4.81	4.17	2.08	.42
3	25.0	14.4	12.5	6.25	1.25
5	41.7	24.0	20.8	10.4	2.08
7.5	62.5	36.1	31.3	15.6	3.13
10	83.3	48.1	41.7	20.8	4.17
15	125.0	72.1	62.5	31.3	6.25
25	208.3	120.2	104.2	52.1	10.4
37.5	312.5	180.3	156.3	78.1	15.6
50	416.7	240.4	208.3	104.2	20.8
75	625.0	360.6	312.5	156.3	31.3
100	833.3	480.8	416.7	208.3	41.7
125	1041.7	601.0	520.8	260.4	52.1
167.5	1395.8	805.3	697.9	349.0	69.8
200	1666.7	961.5	833.3	416.7	83.3
250	2083.3	1201.9	1041.7	520.8	104.2
333	2775.0	1601.0	1387.5	693.8	138.8
500	4166.7	2403.8	2083.3	1041.7	208.3

SIZING THREE-PHASE TRANSFORMERS

To calculate kVA capacity of a 3ϕ transformer

$$kVA_{CAP} = E \times 1.732 \times \frac{I}{1000}$$

Where: kVA_{CAP} = transformer capacity (in kVA)
E = voltage (in Volts)
1.732 = constant (for 3ϕ power)
I = current (in Amps)

kVA Rating	3ϕ FULL-LOAD CURRENTS (AMPS)				
	208 V	240 V	480 V	2400 V	4160 V
3	8.3	7.2	3.6	.72	.416
6	16.7	14.4	7.2	1.44	.83
9	25.0	21.7	10.8	2.17	1.25
15	41.6	36.1	18.0	3.6	2.08
30	83.3	72.2	36.1	7.2	4.16
45	124.9	108.3	54.1	10.8	6.25
75	208.2	180.4	90.2	18.0	10.4
100	277.6	240.6	120.3	24.1	13.9
150	416.4	360.9	180.4	36.1	20.8
225	624.6	541.3	270.6	54.1	31.2
300	832.7	721.7	360.9	72.2	41.6
500	1387.9	1202.8	601.4	120.3	69.4
750	2081.9	1804.3	902.1	180.4	104.1
1000	2775.8	2405.7	1202.8	240.6	138.8
1500	4163.7	3608.5	1804.3	360.9	208.2
2000	5551.6	4811.4	2405.7	481.1	277.6
2500	6939.5	6014.2	3007.1	601.4	347.0
5000	13879.0	12028.5	6014.2	1202.8	694.0
7500	20818.5	18042.7	9021.4	1804.3	1040.9
10000	27758.0	24057.0	12028.5	2405.7	1387.9

TRANSFORMER DERATINGS RESULTING FROM HIGHER AMBIENT TEMPERATURES

Maximum Ambient Temperature (°C) Surrounding Transformer	Maximum Transformer Loading in Percent %
40	100
45	96
50	92
55	88
60	81
65	80
70	76

CALCULATING DERATED kVA CAPACITY FROM HIGHER AMBIENT TEMPERATURES

To calculate the derated kVA capacity of a transformer operating at a higher-than-normal ambient temperature condition, use the formula below:

$$\text{kVA} = \text{rated kVA} \times \text{maximum load}$$

Where: kVA = derated transformer capacity (in kVA)

rated kVA = manufacturer transformer rating (in kVA)

maximum load = maximum transformer loading (in %)

CURRENT-VOLTAGE RELATIONSHIP BETWEEN HIGH SIDE AND LOW SIDE OF A TRANSFORMER

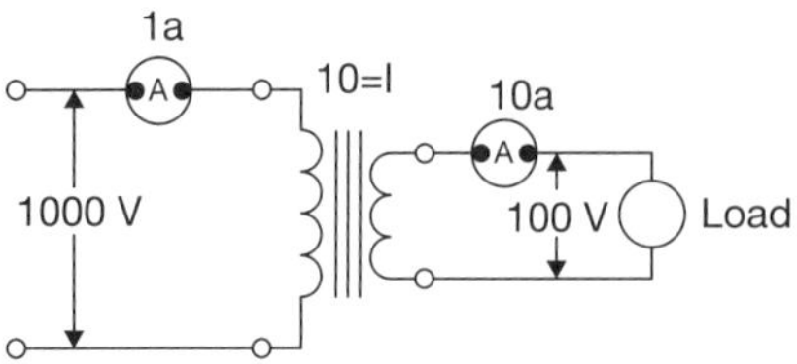

TESTING A TRANSFORMER FOR POLARITY

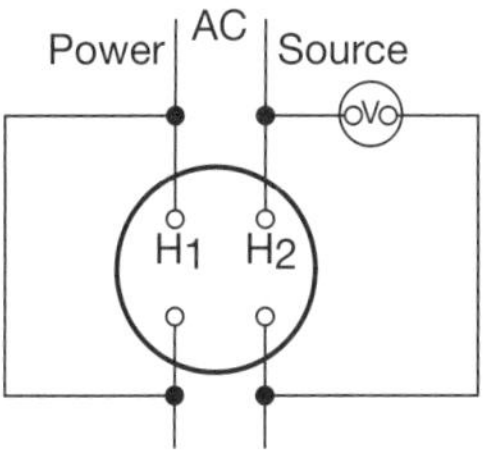

DIAGRAM OF A CURRENT TRANSFORMER

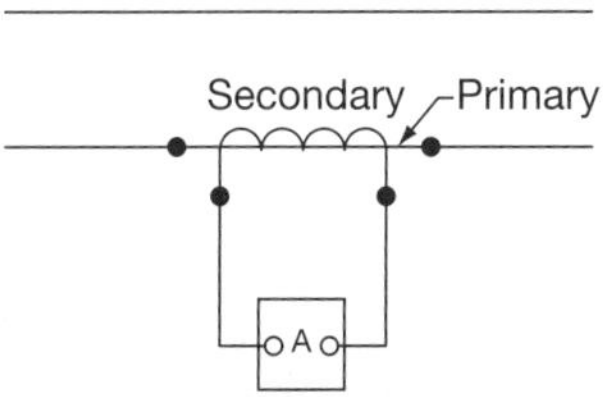

SINGLE-PHASE TRANSFORMER CIRCUITS

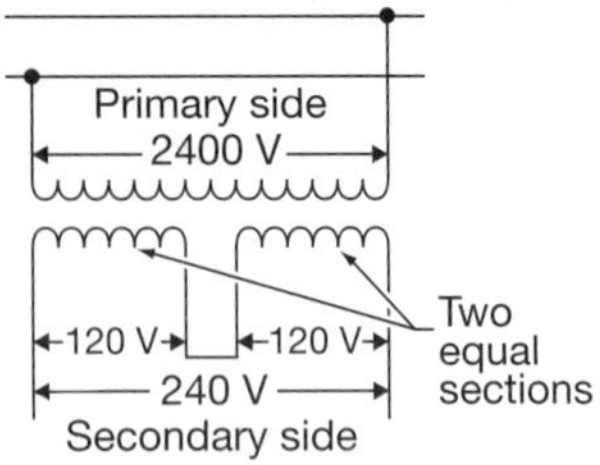

Series-Connected Winding

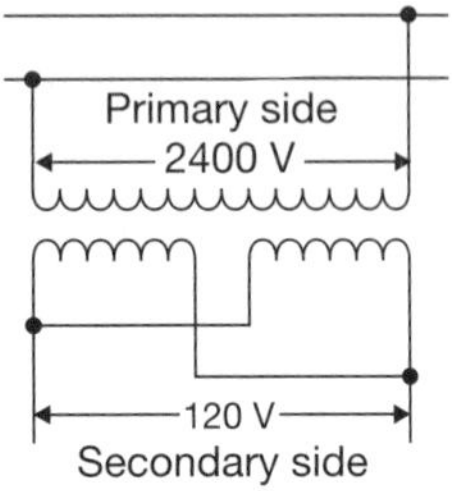

Parallel-Connected Winding

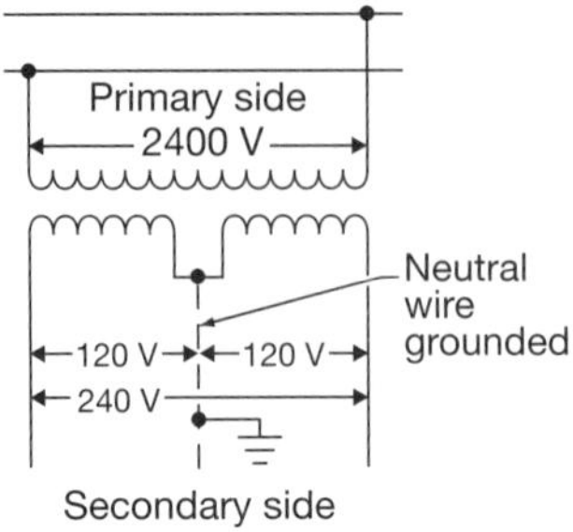

Series-Connected Winding for 3-Wire Service

SINGLE-PHASE TRANSFORMER CONNECTIONS

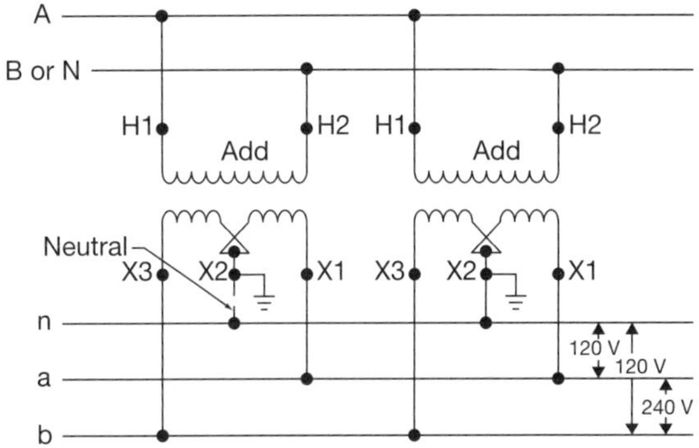

Double-Parallel Connection

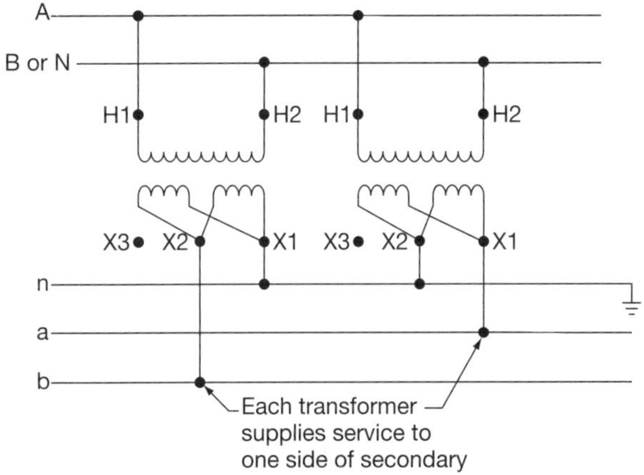

Parallel-Series Secondary Connection

SINGLE-PHASE TRANSFORMER CONNECTIONS *(cont.)*

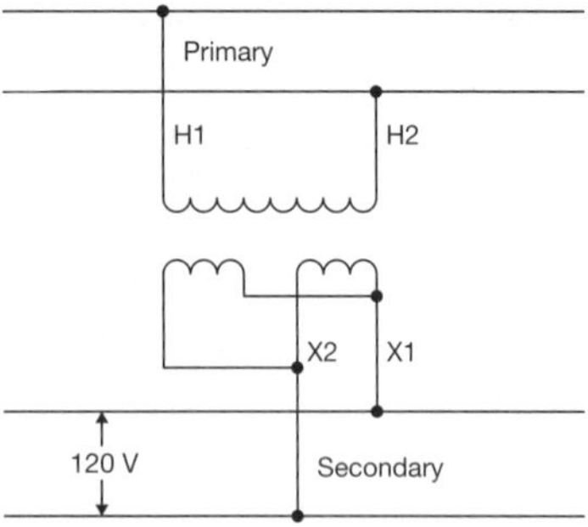

SINGLE-PHASE TO SUPPLY 120 V LIGHTING LOAD

The transformer is connected between high voltage line and load with the 120/240 V winding connected in parallel. This connection is used where the load is comparatively small and the length of the secondary circuit is short. It is often used for a single customer.

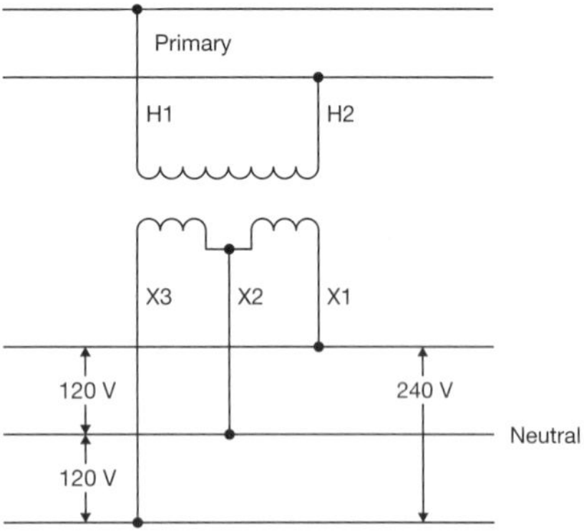

SINGLE-PHASE TO SUPPLY 120/240 V 3-WIRE LIGHTING AND POWER LOAD

Here the 120/240 V winding is connected in series and the mid-point brought out, making it possible to serve both 120 and 240 V loads simultaneously. This connection is used in most urban distribution circuits.

SINGLE-PHASE TRANSFORMER CONNECTIONS *(cont.)*

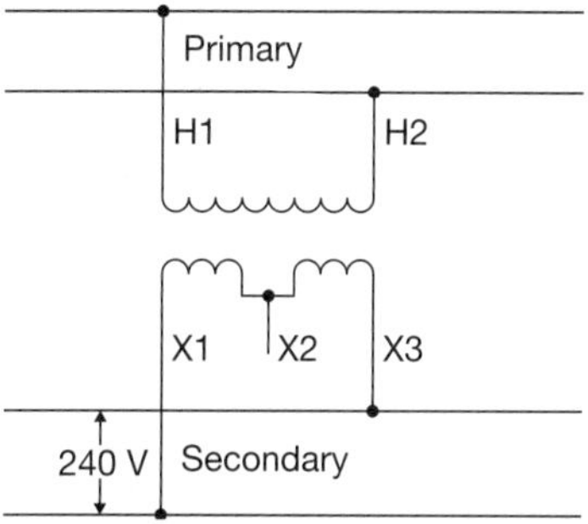

SINGLE-PHASE FOR POWER

In this case the 120/240 V winding is connected in series serving 240 V on a two-wire system. This connection is used for small industrial applications.

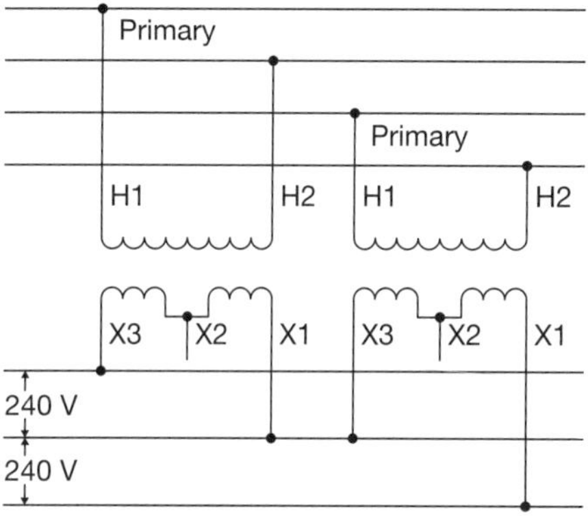

TWO-PHASE CONNECTIONS

This connection consists merely of two single-phase transformers operated 90° out of phase. For a three-wire secondary as shown, the common wire must carry 32 times the load current. In some cases, a four-wire or a five-wire secondary may be used.

VARIOUS TRANSFORMER CONNECTIONS

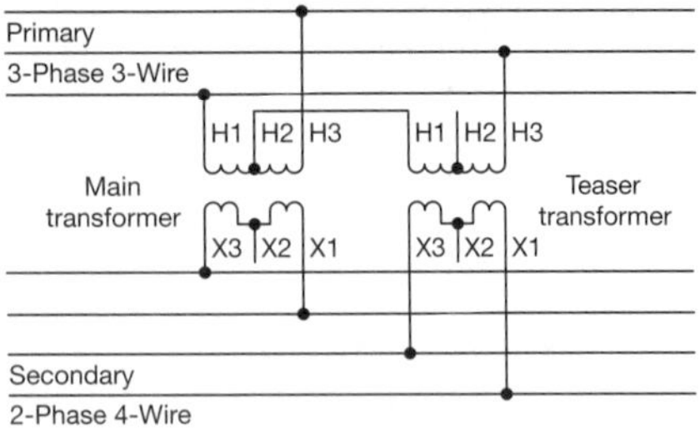

SCOTT CONNECTED THREE-PHASE TO TWO-PHASE

When two-phase power is required from a three-phase system, the Scott connection is used the most. The secondary may be three, four, or five wire. Special taps must be provided at 50% and 86.6% of normal primary voltage to make this connection.

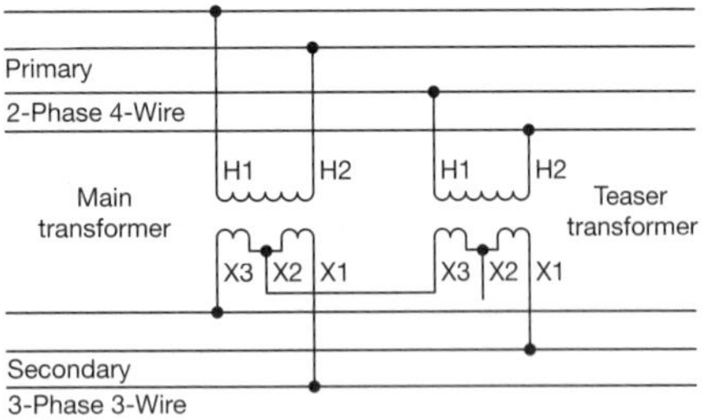

SCOTT CONNECTED TWO-PHASE TO THREE-PHASE

If it should be necessary to supply three-phase power from a two-phase system, the scott connection may be used again. The special taps must be provided on the secondary side, but the connection is similar to the three-phase to two-phase.

To obtain the Scott transformation without a special 86.6% tapped transformer, use one with 10% or two 5% taps to approximate the desired value. A small error of unbalance (overvoltage) occurs that requires care in application.

THREE-PHASE TRANSFORMER CONNECTIONS

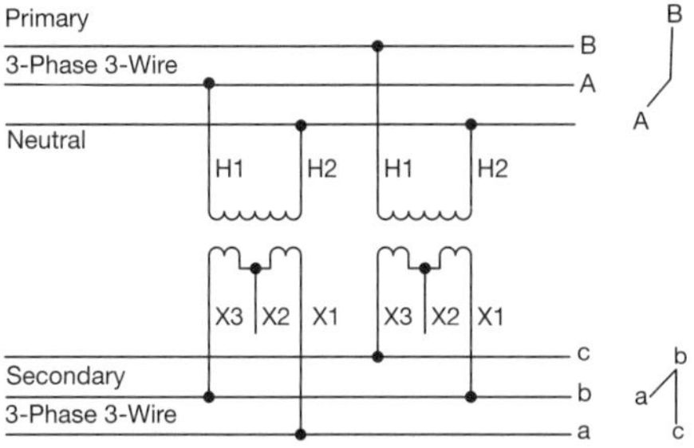

OPEN WYE-DELTA

When operating wye-delta and one phase is disabled, service may be maintained at reduced load as shown. The neutral in this case must be connected to the neutral of the setup bank through a copper conductor. The system is unbalanced, electro-statically and electro-magnetically, so that telephone interference may be expected if the neutral is connected to ground. The useful capacity of the open delta open wye bank is 87% of the capacity of the installed transformers when the two units are identical.

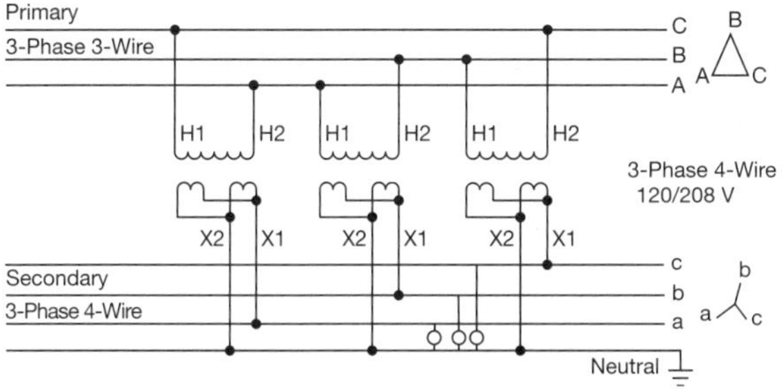

DELTA-WYE FOR LIGHTING AND POWER

In the previous banks the single-phase lighting load is all on one phase resulting in unbalanced primary currents in any one bank. To eliminate this difficulty, the delta-wye system finds many uses. Here the neutral of the secondary three-phase system is grounded and the single-phase loads are connected between the different phase wires and the neutral while the three-phase loads are connected to the phase wires. Thus, the single-phase load can be balanced on three phases in each bank and banks may be paralleled if desired.

THREE-PHASE TRANSFORMER CONNECTIONS *(cont.)*

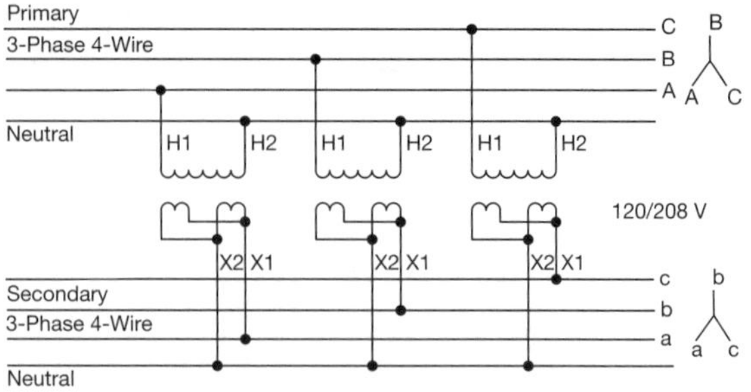

WYE-WYE FOR LIGHTING AND POWER

A system on which the primary voltage was increased from 2400 to 4160 V to increase the potential capacity. The previously delta connected distribution transformers are now connected from line to neutral. The secondaries are connected in wye. The primary neutral is connected to the neutral of the supply voltage through a metallic conductor and carried with the phase conductor to minimize telephone interference. If the neutral of the transformer is isolated from the system neutral an unstable condition results at the transformer neutral caused primarily by third harmonic voltages. If the transformer neutral is connected to ground, the possibility of telephone interference is enhanced and a possibility of resonance between the line capacitance to ground and the magnetizing impedance of the transformer.

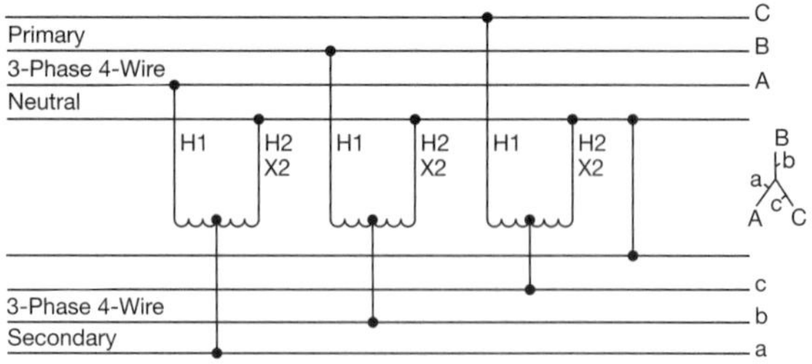

WYE-WYE AUTOTRANSFORMERS FOR SUPPLYING POWER FROM A THREE-PHASE FOUR-WIRE SYSTEM

When ratio of transformation from primary to secondary voltage is small, the best way of stepping down voltage is using autotransformers. It is necessary that the neutral of the autotransformer bank be connected to the system neutral.

THREE-PHASE TRANSFORMER CONNECTIONS *(cont.)*

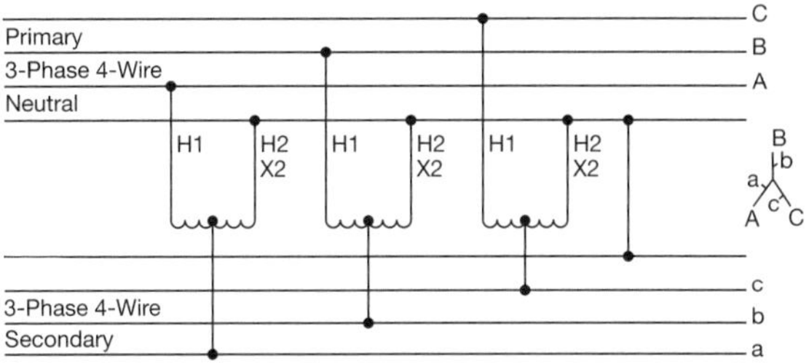

DELTA-DELTA FOR POWER AND LIGHTING

This connection is used to supply a small single-phase lighting load and three-phase power load simultaneously. As shown, the midtap of the secondary of one transformer is grounded. The small lighting load is connected across the transformer with the midtap and the ground wire common to both 120 V circuits. The single-phase lighting load reduces the available three-phase capacity. This requires special watt-hour metering.

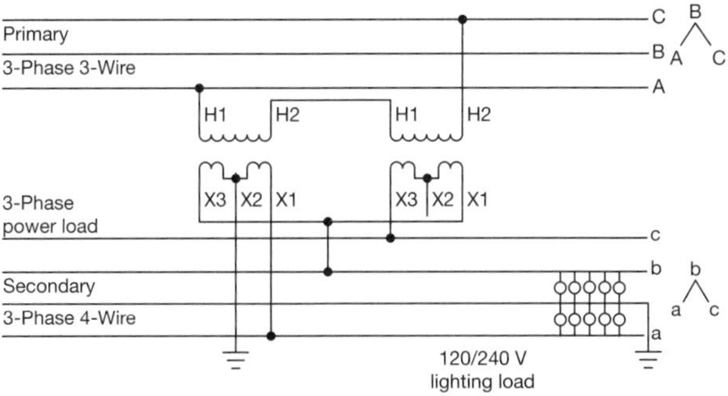

OPEN-DELTA FOR LIGHTING AND POWER

Where the secondary load is a combination of lighting and power, the open-delta connected bank is frequently used. This connection is used when the single-phase lighting load is large as compared with the power load. Here two different size transformers may be used with the lighting load connected across the larger rated unit.

THREE-PHASE TRANSFORMER CONNECTIONS *(cont.)*

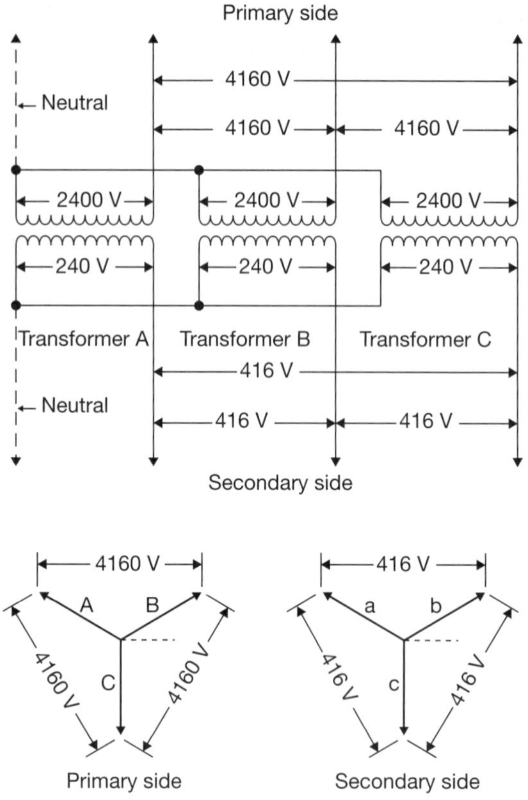

Wye-to-Wye Connection

Three 1ϕ transformers may be connected in a wye-to-wye connection to form a 3ϕ transformer.

THREE-PHASE TRANSFORMER CONNECTIONS *(cont.)*

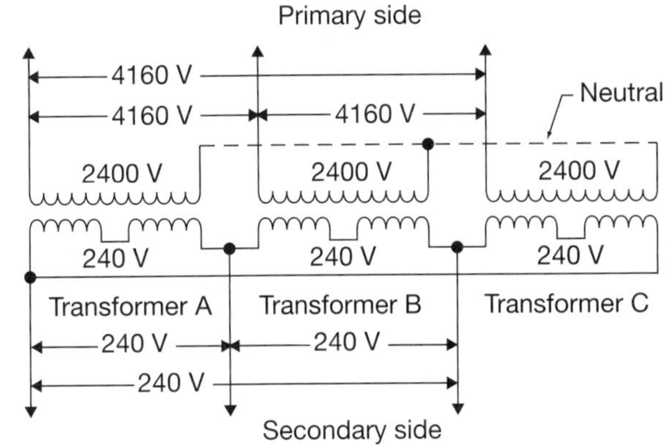

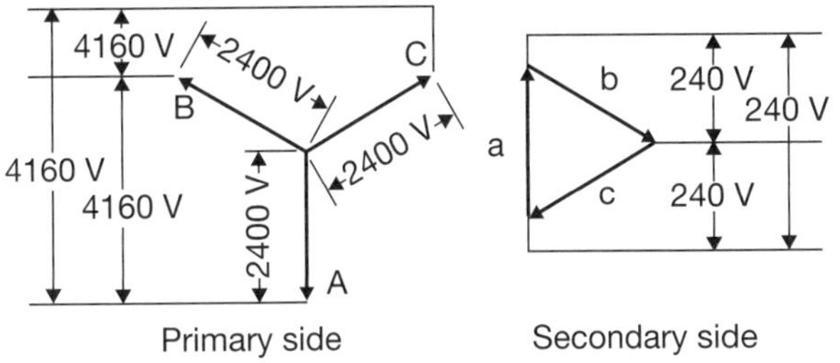

Wye-to-Delta Connection

A wye-to-delta connection permits 1ϕ and 3ϕ loads to be drawn simultaneously from the delta-connected secondary at the same voltage.

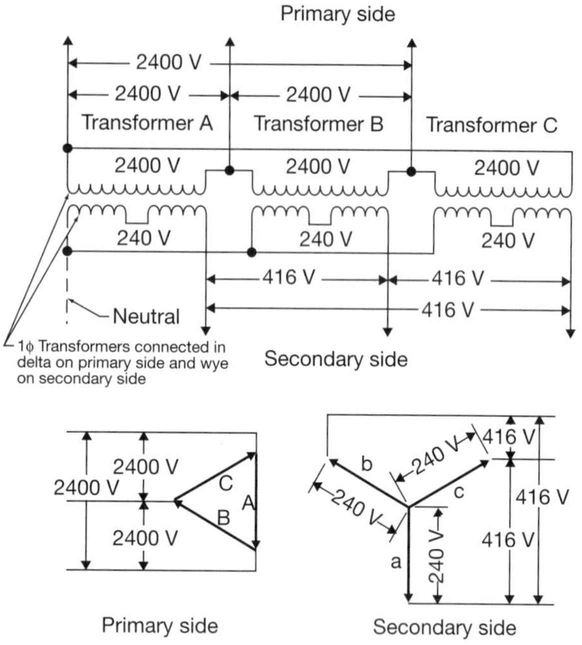

Delta-to-Wye Connection

A 3ϕ delta-to-wye connection is often used for distribution where a four-wire secondary distribution circuit is required.

THREE-PHASE TRANSFORMER CONNECTIONS *(cont.)*

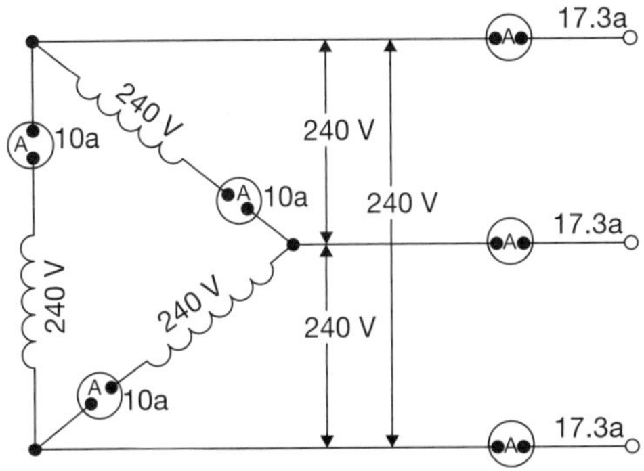

Line current is $\sqrt{3}$ x coil current in a delta connection

Current and voltage values on a delta bank of three single-phase transformer windings. A voltage of 240 volts and a current of 10 amperes are assumed.

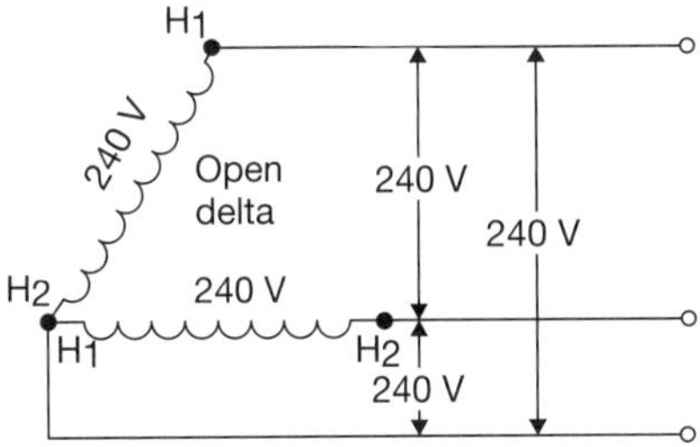

Two Transformers in an Open Delta Arrangement

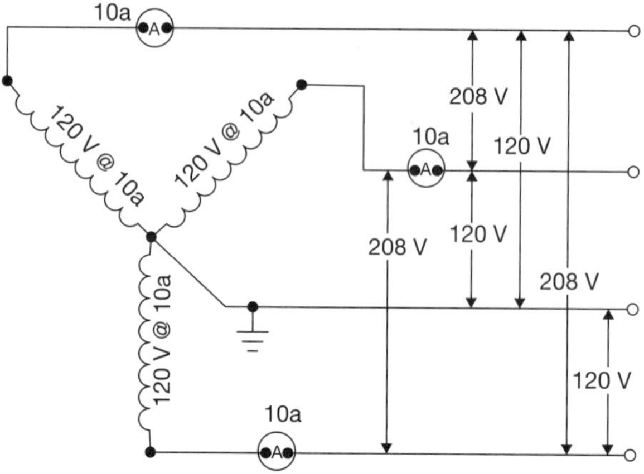

Wye Connection
Line Volts = 1.732 × Coil Volts
Coil Amps = Line Amps

Current and voltage values on a wye bank of three single-phase transformer windings. A voltage of 120 volts and a current of 10 amperes are assumed.

A SPLIT-COIL TRANSFORMER

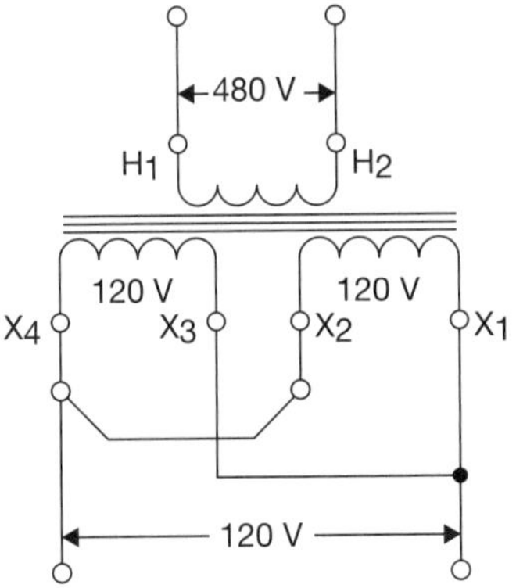

AN ADDITIVE-POLARITY TRANSFORMER

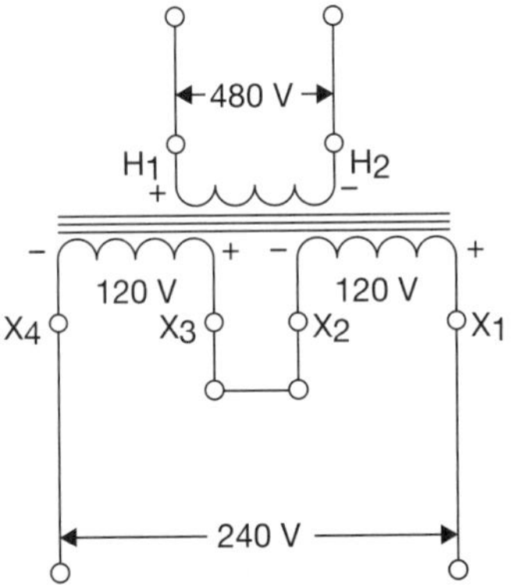

SECONDARY TIES

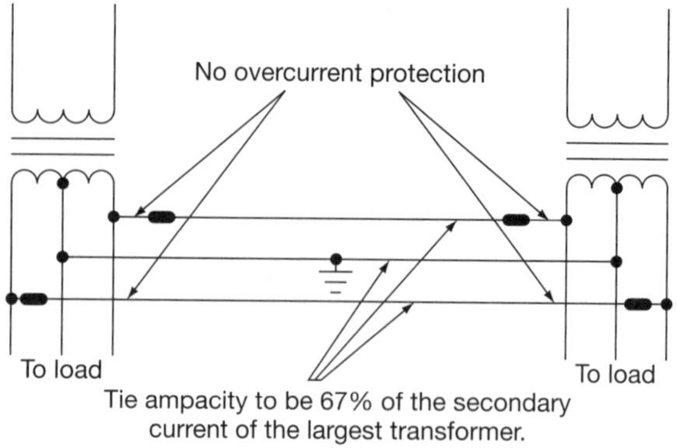

LOADS CONNECTED BETWEEN TRANSFORMER SUPPLY POINTS

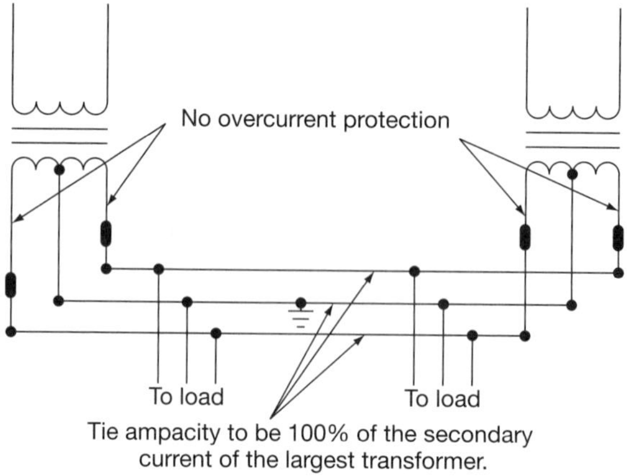

AUTOTRANSFORMER

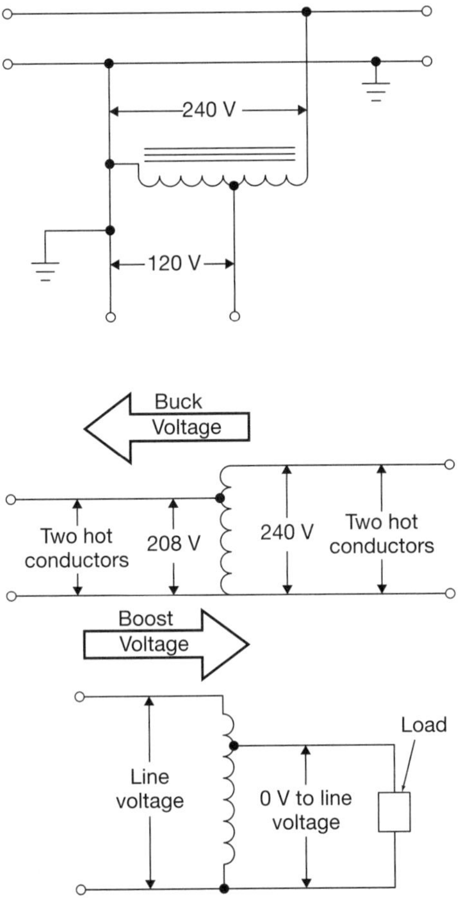

TYPICAL BOOST TRANSFORMER CONNECTIONS

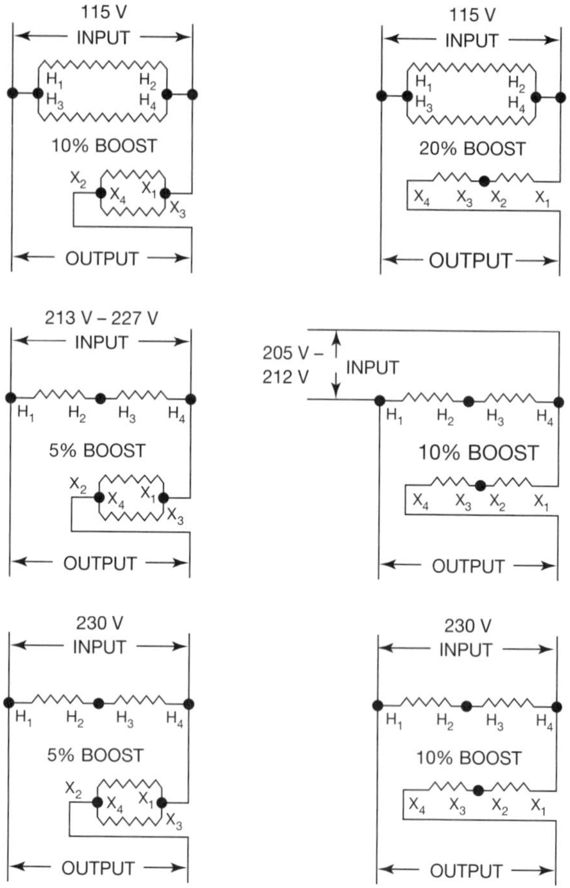

Note: Always verify manufacturer's instructions

TYPICAL BUCK TRANSFORMER CONNECTIONS

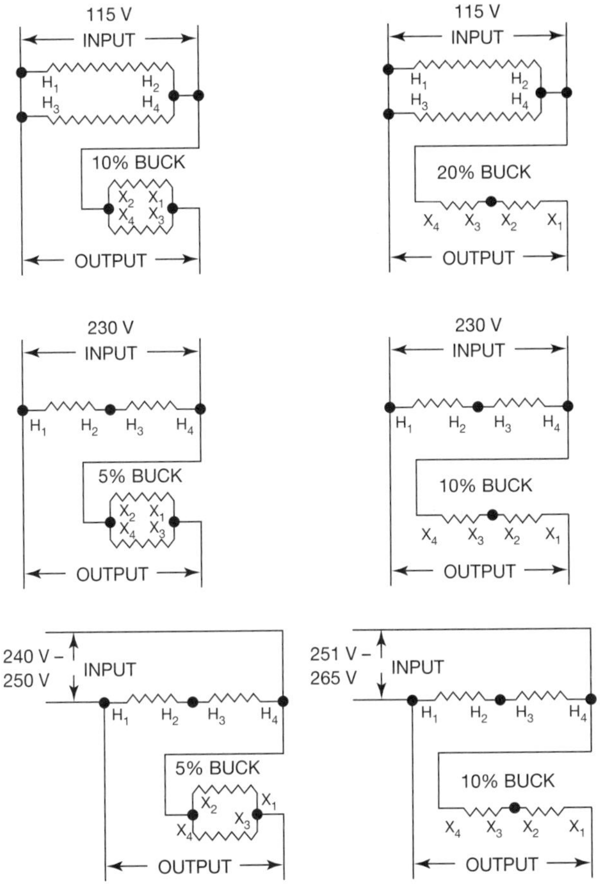

Note: Always verify manufacturer's instructions

TRANSFORMER SOUND LEVELS

kVA Rating	Sound Level (in dB)	Hearing Intensity	Example of Loudness
0–5	40	Barely Audible	Refrigerator running
6–9	40		
10–25	45		
26–50	45		
51–150	50	Quiet	Whisper
151–225	55		
226–300	55		
301–500	60	Moderate	Normal conversation

TRANSFORMING WINDING RATIOS

Tap Winding	Primary Voltage	Secondary Voltage	Ratio
Full Winding	2400	120/240	20:1/10:1
$4\frac{1}{2}$%	2292	120/240	19:1/9.5:1
9%	2184	120/240	18:1/9:1
$13\frac{1}{2}$%	2076	120/240	17:1/8.5:1

SATISFACTORY VOLTAGE LEVELS

Devices	Tolerable Under Voltage	Satisfactory Voltage	Tolerable Over Voltage
Lights, Motors, etc.	112	117	122–127
Stoves, Dryers, etc.	230	235	245–250

CHAPTER 8
Communications and Electronics

TELEPHONE CONNECTIONS			
Typical Inside Wire			
Type of Wire	**Pair No.**	**Pair Color Matches**	
2-pair Wire	1 2	Green Black	Red Yellow
3-pair Wire	1 2 3	White/Blue White/Orange White/Green	Blue/White Orange/White Green/White

Inside Wire Connecting Terminations			
Wire Color		**Wire Function**	
2-pair Wire	3-pair Wire	Service w/o Dial Light	Service with Dial Light
Green Red Black Yellow	White/Blue Blue/White White/Orange Orange/White	Tip Ring Not Used Ground	Tip Ring Transformer Transformer

Typical Fasteners and Recommended Spacing Distances			
Fasteners	**Horizontal**	**Vertical**	**From Corner**
Wire clamp	16.0 in.	16.0 in.	2 in.
Staples (wire)	7.5 in.	7.5 in.	2 in.
Bridle Rings*	4.0 ft.	–	2–8.5 in.*
Drive Rings**	4.0 ft.	8.0 ft.	2–8.5 in.*

*When changing direction space fasteners to hold wire at 45-degree angle.
**To avoid injury do not use drive rings below a 6 foot clearance, use bridle rings.

STANDARD TELECOM COLOR CODING		
PAIR #	TIP (+) COLOR	RING (–) COLOR
1	White	Blue
2	White	Orange
3	White	Green
4	White	Brown
5	White	Slate
6	Red	Blue
7	Red	Orange
8	Red	Green
9	Red	Brown
10	Red	Slate
11	Black	Blue
12	Black	Orange
13	Black	Green
14	Black	Brown
15	Black	Slate
16	Yellow	Blue
17	Yellow	Orange
18	Yellow	Green
19	Yellow	Brown
20	Yellow	Slate
21	Violet	Blue
22	Violet	Orange
23	Violet	Green
24	Violet	Brown
25	Violet	Slate

SEPARATION AND PHYSICAL PROTECTION FOR PREMISES WIRING

Table applies only to telephone wiring from the Network Interface modular jacks to telephone equipment. Separations apply to crossing and to parallel runs (minimum separations).

Types of Wire Involved		Minimum Separations	Alternatives
Electric Supply	Bare light or power wire of any voltage	5 ft.	No Alternative
	Open wiring not over 300 V	2 in.	See Note 1.
Radio & TV	Antenna lead-in and grds.	4 in.	See Note 1.
Signal or Control Wires	Open wiring or wires in conduit or cable	None	N/A
Comm. Wires	Coaxial cables w/shield	None	N/A
Telephone Drop Wire	Using fused protectors	2 in.	See Note 1.
	Using fuseless protectors		None
Sign	Neon Signs and wiring		6 in.
Lightning Sys.	Lightning rods and wires		6 ft.

Note 1: If minimum separations cannot be obtained, additional protection required.

MINIMUM SEPARATION DISTANCE FROM POWER SOURCE AT 480 V OR LESS

CONDITION	<2 kVA	2-5 kVA	>5 kVA
Unshielded power lines or equipment near open or non-metal pathways	5 in.	12 in.	24 in.
Unshielded power lines or equipment near a grounded metal conduit pathway	2.5 in.	6 in.	12 in.
Power lines enclosed in a grounded metal conduit (or equivalent)	—	6 in.	12 in.
Transformers and electric motors	40 in.	40 in.	40 in.
Fluorescent lighting	12 in.	12 in.	12 in.

CONDUCTORS ENTERING BUILDINGS

If on the same pole, or run parallel in span:

1. Communications conductors must be located below power conductors.
2. Communications conductors cannot be connected to crossarms.
3. Power drops must be separated from comm. drops by 12 inches.

Above roofs, communications conductors must have the following clearances:

1. Flat roofs: 8 feet.
2. Garages and other auxiliary buildings: None required.
3. Overhangs, where no more than 4 feet of communications cable will run over the area: 18 inches.
4. Where the roof slope is 4 inches rise for every 12 inches horizontally: 3 feet.

Underground communications conductors must be separated from power conductors in manhole or handholes by brick, concrete, or tile partitions.

Communications conductors should be kept at least 6 feet away from lightning protection system conductors.

CIRCUIT PROTECTION REQUIREMENTS

1. **Protectors** are surge arresters and are required for all aerial circuits not confined with a *block*. They must be installed on all circults with a block that could accidentally contact power circuits over 300 volts to ground.
2. **Metal sheaths** of any communications cables must be grounded or interrupted with an insulating joint as close as practicable to the point where they enter any building (where the cable emerges through an exterior wall or concrete floor slab, or from a grounded rigid or intermediate metal conduit).
3. **Grounding conductors** for communications circuits must be copper or some other corrosion-resistant material, and have insulation suitable for the area in which it is installed.
4. **Grounding conductors** may be no smaller than 14 AWG.
5. The **grounding conductor** must be run as directly as possible to the grounding electrode, and be protected if necessary.
6. If the **grounding conductor** Is protected by metal raceway, it must be bonded to the grounding conductor on both ends.

CIRCUIT PROTECTION REQUIREMENTS *(cont.)*

Grounding electrodes for communications ground may be any of the following:

1. The grounding electrode of an electrical power system.
2. A grounded interior metal piping system.
3. Metal power service raceway.
4. Power service equipment enclosures.
5. A separate grounding electrode.
6. Any acceptable power system grounding electrode.
7. A grounded metal structure.
8. A ground rod or pipe at least 5 feet long and 1/2 inch in diameter. This rod should be driven into damp (if possible) earth, and kept separate.

If the power and communications systems use separate grounding electrodes, bond together with 6 AWG copper. Other electrodes may be bonded also.

For mobile homes, if there is no service equipment or disconnect within 30 feet of the mobile home wall, the communications circuit must have its own grounding electrode. In this case, or if the mobile home is connected with cord and plug, the communications circuit protector must be bonded to the mobile home frame or grounding terminal with a copper conductor no smaller than 12 AWG.

INTERIOR COMMUNICATIONS CONDUCTORS

Must be kept at least 2 inches away from power or Class 1 conductors, unless permanently separated from them or they are enclosed in one of the following:

1. Raceway.
2. Type AC, MC, UF, or NM cable, or metal-sheathed cable.

Communications cables are allowed in the same raceway, box, or cable with:

1. Class 2 and 3 remote-control, signaling, and power-limited circuits.
2. Power-limited fire protective signaling systems.
3. Conductive or nonconductive optical fiber cables.
4. Community antenna television and radio distribution systems.

Communications conductors are not allowed to be in the same raceway or fitting with power or Class 1 circuits.

Communications conductors are not allowed to be supported by raceways unless the raceway runs directly to the piece of equipment the communications circuit serves.

TELEPHONE CONNECTIONS

Telephone cable contains four wires, colored green, red, black, and yellow. A one line telephone requires only two wires to operate. Green and red are the conductors used. In a four-wire modular connector, the green and red conductors are in the inside positions, the black and yellow wires in the outer positions. For two-line phones, the inside wires (green and red) carry line 1, and the outside wires (black and yellow) carry line 2.

EIA COLOR CODE

Pair 1 – White/Blue (white with blue stripe) and Blue
Pair 2 – White/Orange and Orange
Pair 3 – White/Green and Green
Pair 4 – White/Brown and Brown

TWISTED-PAIR PLUGS AND JACKS

There are three major types of twisted-pair jacks:

RJ-type connectors (phone plugs)
Pin-connectors
Genderless connectors (IBM sexless data connectors)

The RJ-type (registered jack) refers to the standard used for most telephone jacks. Pin-connector refers to twisted pair such as the RS-232 connector, which have male and female pin receptacles. Genderless connectors have no separate male or female component.

PHONE JACKS

A **four-conductor jack**, supporting two twisted-pairs, is the one used for connecting telephone handsets to receivers.

A **six-conductor jack**, supporting three twisted-pairs, is the RJ-11 jack used to connect telephones to the telephone company. An **eight-conductor jack** is the R-45 jack, which is intended for use as the user-site interface for ISDN terminals.

These types of jacks are often keyed, so that the wrong type of plug cannot be inserted into the jack.

CROSS CONNECTIONS

Cross connections are made at **terminal blocks**. A block is typically a rectangular, white plastic unit, with metal connection points. The most common type is called a punchdown block. Connections are made by pushing the insulated wires into their places. When "punched" down, the connector cuts through the insulation, and makes the appropriate connection.

Connections are made between punch-down blocks by using **patch cords**, which are short lengths of cable that can be terminated into the punch-down slots, or that are equipped with connectors on each end.

When different systems must be connected together, cross-connects are used.

CATEGORY CABLING

Category 1 cable is the old standard type of telephone cable, with four conductors.
Category 2 Obsolete.
Category 3 cable is used for digital voice and data transmission rates up to 10 Mbit/s (Megabits per second). Common types of data transmission over this communications cable would be UTP Token Ring (4 Mbit/s) and 10Base-T (10 Mbit/s).
Category 4 Obsolete.
Category 5 cable is used for sending voice and data at speeds up to 100Mbit/s.
Category 6 cable is used for sending data at speeds up to 200 or 250Mbit/s.

INSTALLATION REQUIREMENTS

Article 800 of the NEC® covers communication circuits, such as telephone systems and outside wiring for fire and burglar alarm systems. Generally these circults must be separated from power circuits and grounded. In addition, all such circuits that run out of doors (even if only partially) must be provided with circuit protectors (surge or voltage suppressors).

TYPES OF DATA NETWORKS

10Base2	Ethernet running over thin, 50 Ohm baseband coaxial cable at 10 Mbit/s. 10Base2 is also commonly referred to as thin-Ethernet.
10BaseT	Ethernet over twisted-pair at 10Mbit/s
100BaseTX	Ethernet over twisted-pair at 100Mbit/s
1000BaseT	Ethernet over twisted-pair at 1Gbit/s
10GBaseT	Ethernet over twisted-pair at 10Gbit/s
10GBaseR	Ethernet over 50 micron fiber at 10Gbit/s

STANDARD CONFIGURATIONS FOR SEVERAL COMMON NETWORKS

ATM 155 Mbps	pairs 2 and 4 (pins 1-2, 7-8)
Ethernet 10BaseT	pairs 2 and 3 (pins 1-2, 3-6)
Ethernet 100BaseT4	pairs 2 and 3 (4T+) (pins 1-2, 3-6)
Ethernet 100BaseT8	pairs 1, 2, 3 and 4 (pins 4-5, 1-2, 3-6, 7-8)
Token-Ring	pairs 1 and 3 (pins 4-5, 3-6)
TP-PMD	pairs 2 and 4 (pins 1-2, 7-8)
100VG-AnyLAN	pairs 1, 2, 3 and 4 (pins 4-5, 1-2, 3-6, 7-8)

ETHERNET 10BaseT STRAIGHT-THRU PATCH CORD

RJ45 Plug	RJ45 Plug	
T2 White/Orange	1	TxData +
R2 Orange	2	TxData −
T3 White/Green	3	RecvData +
R1 Blue	4	
T1 White/Blue	5	
R3 Green	6	RecvData −
T4 White/Brown	7	
R4 Brown	8	

ETHERNET 10BaseT CROSSOVER PATCH CORD

This cable is used to cascade hubs, or for connecting two Ethernet stations back-to-back without a hub.

RJ45 Plug	RJ45 Plug
1 Tx+	Rx+ 3
2 Tx−	Rx− 6
3 Rx+	Tx+ 1
6 Rx−	Tx− 2

EIA-606 COLORS AND FUNCTIONS OF CABLING

Blue	Horizontal voice cables
Brown	Inter-building backbone
Gray	Second-level backbone
Green	Network connections and auxiliary circuits
Orange	Demarcation point
Purple	First-level backbone
Red	Key-type telephone systems
Silver or White	Horizontal data cables, computer and PBX
Yellow	Auxiliary, maintenance and security alarms

THERMOSTAT WIRING GUIDE

- Red wire to the white wire turns furnace on.
- Red wire to the green wire turns fan on.
- Red wire to the yellow wire turns A/C on.

or

- **White Wire = Heat On**
- **Green Wire = Fan On**
- **Yellow Wire = A/C On**

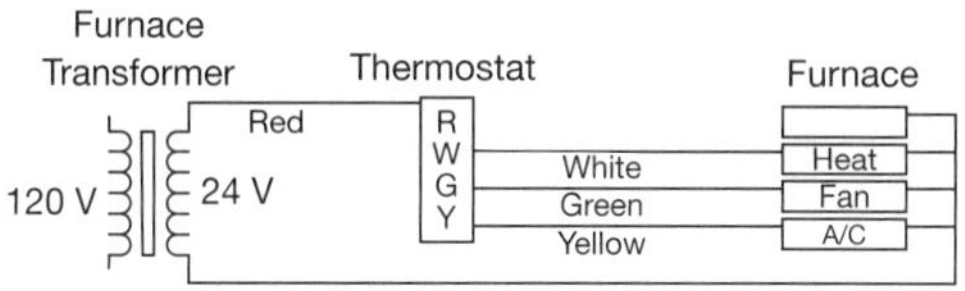

WIRING GUIDE FOR THERMOSTAT, FURNACE AND AIR CONDITIONING

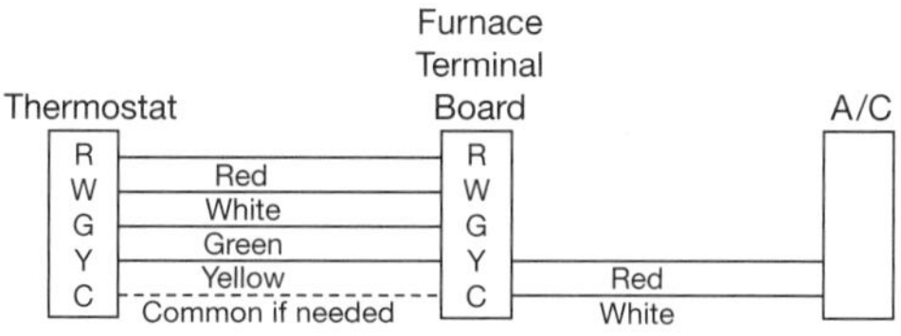

For thermostats with a 2-stage setup, for 2 furnaces and 2 compressors, remove the jumper between RC and RH. For 1-stage, leave the jumper in place.

BASIC THERMOSTAT CIRCUIT

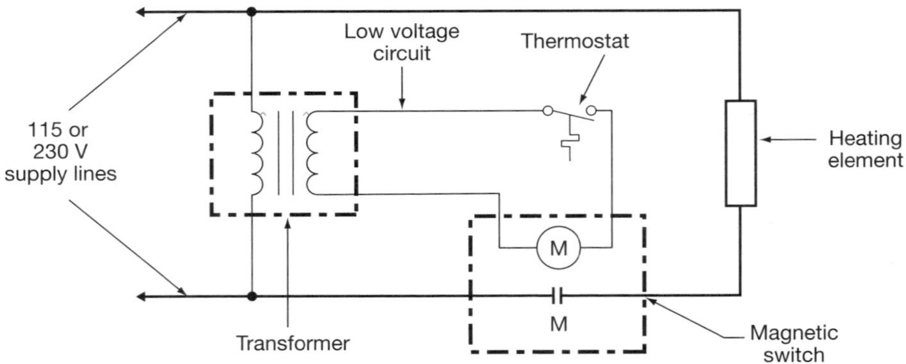

TWIN-TYPE THERMOSTAT CIRCUIT

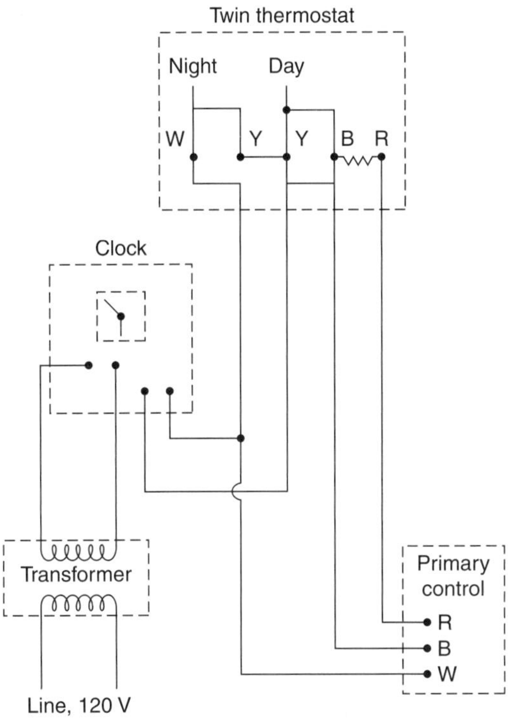

INTRUDER ALARM CIRCUIT WITH ANNUNCIATOR AND RELAY

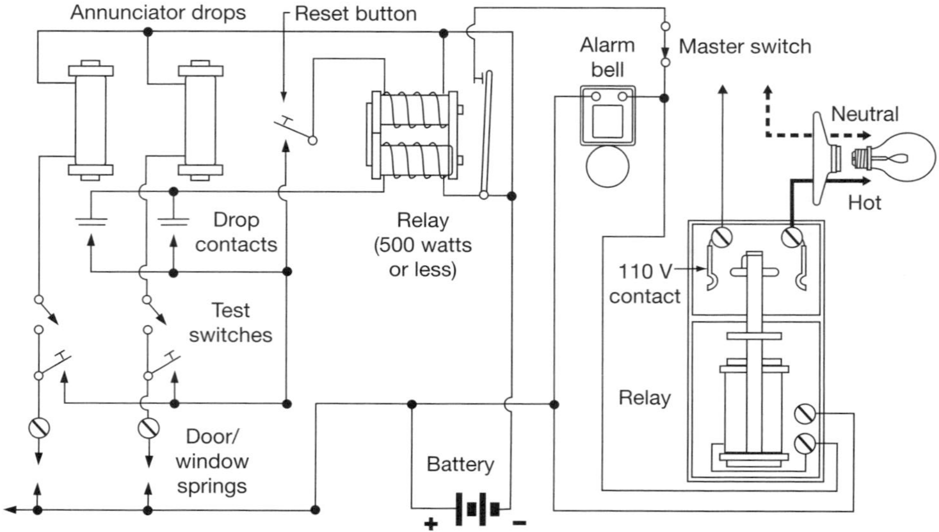

ELECTRONIC WIRING COLOR CODES

Electronic Applications

Color	Circuit Type
Black	Chassis grounds, returns, primary leads
Blue	Plate leads, transistor collectors, FET drain
Brown	Filaments, plate start lead
Gray	AC main power leads
Green	Transistor base, finish grid, diodes, FET gate
Orange	Transistor base 2, screen grid
Red	B plus dc power supply
Violet	Power supply minus
White	B – C minus of bias supply, AVC – AGC return
Yellow	Emitters-cathode and transistor, FET source

Stereo Audio Channels

Color	Circuit Type
White	Left channel high side
Blue	Left channel low side
Red	Right channel high side
Green	Right channel low side

AF Transformers (Audio)

Color	Circuit Type
Black	Ground line
Blue	Plate, collector, or drain lead; End of primary winding
Brown	Start primary loop; Opposite to blue lead
Green	High side, end secondary loop
Red	B plus, center tap push-pull loop
Yellow	Secondary center tap

IF Transformers (Intermediate Frequency)

Color	Circuit Type
Blue	Primary high side of plate, collector, or drain lead
Green	Secondary high side for output
Red	Low side of primary returning B plus
Violet	Secondary outputs
White	Secondary low side

COLOR CODES FOR RESISTORS

Radial Lead Resistor

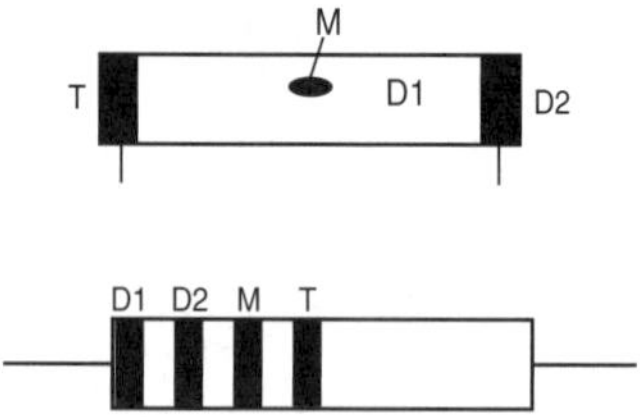

Axial Lead Resistor

Color	1st # (D1)	2nd # (D2)	Multiplier (M)	Tolerance (T)
No Color	–	–	–	20%
Silver	–	–	0.01	10%
Gold	–	–	0.1	5%
White	9	9	10^9	–
Gray	8	8	100,000,000	–
Violet	7	7	10,000,000	–
Blue	6	6	1,000,000	–
Green	5	5	100,000	–
Yellow	4	4	10,000	–
Orange	3	3	1,000	4%
Red	2	2	100	3%
Brown	1	1	10	2%
Black	0	0	1	1%

Example: Yellow – Blue – Brown – Silver = 460 ohms

If only Band D1 is wide, it indicates that the resistor is wirewound. If Band D1 is wide and there is also a blue fifth band to the right of Band T on the Axial Lead Resistor, it indicates the resistor is wirewound and flame proof.

STANDARD VALUES FOR RESISTORS

Values for 5% class   k = kilohms = 1,000 ohms   m = megohms = 1,000,000 ohms

1	8.2	68	560	4.7k	39k	330k	2.7m
1.1	9.1	75	620	5.1k	43k	360k	3.0m
1.2	10	82	680	5.6k	47k	390k	3.3m
1.3	11	91	750	6.2k	51k	430k	3.6m
1.5	12	100	820	6.8k	56k	470k	3.9m
1.6	13	110	910	7.5k	62k	510k	4.3m
1.8	15	120	1.0k	8.2k	68k	560k	4.7m
2.0	16	130	1.1k	9.1k	75k	620k	5.1m
2.2	18	150	1.2k	10k	82k	680k	5.6m
2.4	20	160	1.3k	11k	91k	750k	6.2m
2.7	22	180	1.5k	12k	100k	820k	6.8m
3.0	24	200	1.6k	13k	110k	910k	7.5m
3.3	28	220	1.8k	15k	120k	1.0m	8.2m
3.6	30	240	2.0k	16k	130k	1.1m	9.1m
3.9	33	270	2.2k	18k	150k	1.2m	10.0m
4.3	36	300	2.4k	20k	160k	1.3m	
4.7	39	330	2.7k	22k	180k	1.5m	
5.1	43	360	3.0k	24k	200k	1.6m	
5.6	47	390	3.3k	27k	220k	1.8m	
6.2	51	430	3.6k	30k	240k	2.0m	
6.8	56	470	3.9k	33k	270k	2.2m	
7.5	62	510	4.3k	36k	300k	2.4m	

CAPACITORS

Mica Capacitor

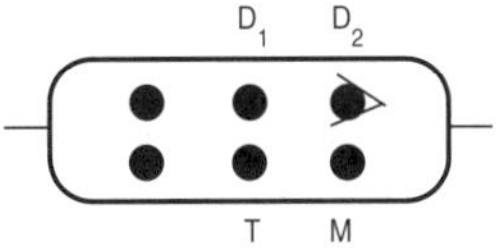

Disc Capacitor

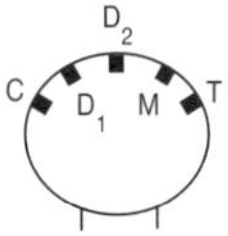

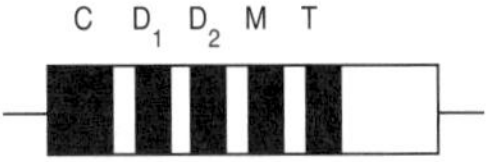

Ceramic Capacitor

Ceramic disc capacitors are normally labeled. If the number is less than 1 then the value is in picofarads, if greater than 1 the value is in microfarads. The letter R can be used in place of a decimal; for example, 2R9=2.9

COLOR CODES FOR CAPACITORS

Color	1st # (D1)	2nd # (D2)	Multiplier (M)	Tolerance (T)
No Color	–	–	–	20%
Silver	–	–	0.01	10%
Gold	–	–	0.1	5%
White	9	9	10^9	9%
Gray	8	8	100,000,000	8%
Violet	7	7	10,000,000	7%
Blue	6	6	1,000,000	6%
Green	5	5	100,000	5%
Yellow	4	4	10,000	4%
Orange	3	3	1,000	3%
Red	2	2	100	2%
Brown	1	1	10	1%
Black	0	0	1	20%

COLOR CODES FOR CERAMIC CAPACITORS

Color	Tolerance (T)		Decimal Multiplier (M)	Temp Coef ppm/°C (C)
	Above 10pf	Below 10pf		
White	10	1.0	0.1	500
Gray	–	0.25	0.01	30
Violet	–	–	–	−750
Blue	–	–	–	−470
Green	5	0.5	–	−330
Yellow	–	–	–	−220
Orange	–	–	1000	−150
Red	2	–	100	−80
Brown	1	–	10	−30
Black	20	2.0	1	0

CHARACTERISTICS OF LEAD – ACID BATTERIES

TEMPERATURE (F) VERSUS BATTERY EFFICIENCY (%)

–20° = 18%	20° = 58%
–10° = 33%	30° = 64%
0° = 40%	50° = 82%
10° = 50%	80° = 100%

CHARGE	SPECIFIC GRAVITY OF ACID
Discharged	1.11 to 1.12
Very Low Capacity	1.13 to 1.15
25% of Capacity	1.15 to 1.17
50% of Capacity	1.20 to 1.22
75% of Capacity	1.24 to 1.26
100% of Capacity	1.26 to 1.28
Overcharged	1.30 to 1.32

MAGNETIC PERMEABILITY OF SOME COMMON MATERIALS

Substance	Permeability (approx.)
Aluminum	Slightly more than 1
Bismuth	Slightly less than 1
Cobalt	60–70
Ferrite	100–300
Free space	1
Iron	60–100
Iron, refined	3000–8000
Nickel	50–60
Permalloy	3000–30,000
Silver	Slightly less than 1
Steel	300–600
Super permalloys	100,000–1,000,000
Wax	Slightly less than 1
Wood, dry	Slightly less than 1

TRANSISTOR CIRCUIT ABBREVIATIONS

Quantity	Abbreviations
Base-emitter voltage	E_B, V_B, E_{BE}, V_{BE}
Collector-emitter voltage	E_C, V_C, E_{CE}, V_{CE}
Collector-base voltage	E_{BC}, V_{BC}, $E_{CB}V_{CB}$
Gate-source voltage	E_G, V_G, E_{GS}, V_{GS}
Drain-source voltage	E_D, V_D, E_{DS}, V_{DS}
Drain-gate voltage	E_{DG}, V_{DG}, E_{DG}, V_{DG}
Emitter current	I_E
Base current	I_B, I_{BE}, I_{EB}
Collector current	I_C, I_{CE}, I_{EC}
Source current	I_S
Gate current	I_G, I_{GS}, I_{SG}*
Drain current	I_D, I_{DS}, I_{SD}

*This is almost always significant.

RADIO FREQUENCY CLASSIFICATIONS

Classification	Abbreviation	Frequency range
Very Low Frequency	VLF	9 kHz and below
Low Frequency (Longwave)	LF	30 kHz - 300 kHz
Medium Frequency	MF	300 kHz - 2MHz
High Frequency (Shortwave)	HF	3 MHz - 30MHz
Very High Frequency	VHF	30 MHz - 300 MHz
Ultra High Frequency	UHF	300 MHz - 3 GHz
Microwaves	MW	3 GHz and more

SMALL TUBE FUSES

CHARACTERISTICS	Fuse Type	Fuse Diameter	Fuse Length
Dual Element, time delay, glass tube	MDL	0.250"	1.250"
Dual Element, glass tube	MDX	0.250"	1.250"
Dual Element, glass tube, pigtail	MDV	0.250"	1.250"
Ceramic Body, normal, 200% 15 sec	3AB	0.250"	1.250"
Metric, fast blow, high int., 210% 30 minutes	216	5mm	20mm
Glass, metric, fast blow, 210% 30 minutes	217	5mm	20mm
Glass, metric, slow blow, 210% 2 minutes	218	5mm	20mm
No Delay, ceramic, 110% rating, opens at 135% load in one hour	ABC	0.250"	1.250"
Fast Acting, glass tube, 110% rating, opens at 135% load in one hour	AGC	0.250"	1.250"
Fast Acting, glass tube	AGX	0.250"	1.000"
No Delay, 200% 15 sec	BLF	0.406"	1.500"
No Delay, military, 200% 15 sec	BLN	0.406"	1.500"
Fast Cleaning, 600V, 135% 1 hr	BLS	0.406"	1.375"
Time Delay, indicator pin, 135% 1 hr	FLA	0.406"	1.500"
Dual Element, delay, 200% 12 sec	FLM	0.406"	1.500"
Dual Element, delay, 500V, 200% 12 sec	FLQ	0.406"	1.500"
Slow Blow, time delay	FNM	0.406"	1.250"
Slow Blow, indicator, metal pin pops outs indicating blown, dual element	FNA	0.406"	1.500"
Rectifier Fuse, fast, low let through	GBB	0.250"	1.250"
Indicator Fuse, metal pin pops out indicating blown, 110% rating	GLD	0.250"	1.250"
Metric, fast acting	GGS	5mm	20mm
Fast, current limiting, 600V, 135% 1 hr	KLK	0.406"	1.500"
Fast, protect solid state, 250% 1 sec	KLW	0.406"	1.500"
Slow Blow, time delay size rejection also	SC	0.406"	1.625" to 2.250"
Slow Blow, glass body, 200% 5 sec	218000	0.197"	0.787"
Slow Blow, glass body, 200% 5 sec	313000	0.250"	1.250"
Slow Blow, ceramic, 200% 5 sec	326000	0.250"	1.250"
Auto Glass, fast blow, 200% 5 sec	1AG	0.250"	0.625"
Auto Glass, fast blow, 200% 10 sec	2AG	0.177"	0.570"
Auto Glass, fast blow, 200% 5 sec	3AG	0.250"	1.250"
Auto Glass, fast blow, 200% 5 sec	4AG	0.281"	1.250"
Auto Glass, fast blow, 200% 5 sec	8AG	0.406"	1.500"
Auto Glass, fast blow, 200% 5 sec	7AG	0.250"	0.875"
Auto Glass, fast blow, 200% 5 sec	8AG	0.250"	1.000"
Auto Glass, fast blow, 200% 5 sec	9AG	0.250"	1.438"

The percentage and time figures mean that a 135% overload will blow a KLK type fuse in 1 hour. (example).

CHAPTER 9
Electrical and Job Safety

ELECTRICAL SAFETY GUIDELINES
• Always comply with the NEC®. • Use UL-approved appliances, components and equipment. • Keep electrical grounding circuits in good condition. Ground any conductive component or element that does not have to be energized. The grounding connection must be a low-resistance conductor heavy enough to carry the largest fault current that may occur. • Turn OFF, lock out and tag disconnect switches when working on any electrical circuit or equipment. Test all circuits after they are turned OFF. • Use double-insulated power tools or power tools that include a third conductor grounding terminal, which provides a path for fault current. • Always use protective and safety equipment. • Check conductors, cords, components and equipment for signs of wear or damage. • Never throw water on an electrical fire. Turn OFF the power and use a Class C–rated fire extinguisher. • Never work alone when working in a dangerous area or with dangerous equipment. • Learn CPR and first aid. • Do not work in poorly lighted areas.

ELECTRICAL SAFETY GUIDELINES *(cont.)*

- Always use nonconductive ladders, never metal.
- Ensure there are no atmospheric hazards such as flammable dust or vapor in the work area.
- Use one hand when working on a live circuit to reduce the chance of an electrical shock passing through the heart and lungs.
- Never bypass or disable fuses or circuit breakers.
- An effective lockout/tagout system is in place.
- Frayed, damaged or worn electrical cords or cables are promptly replaced.

EFFECT OF ELECTRIC CURRENT ON THE HUMAN BODY

Current (ma)	Effect	Result
0 to 6	Slight sensation possible	None
6 to 15	Painful shock muscular contraction	Possible No "Let Go"
15 to 20	Painful shock Frozen until circuit is de-energized	No "Let Go"
20 to 50	Severe muscular contractions Asphyxia	Often Fatal
50 to 200	Ventricular fibrillation	Probably Fatal
200 +	Heart movement stops	Fatal

ELECTRICAL SAFETY CHECKLIST

- GFCI protected receptacles, cords, etc.
- Work on new and existing energized (hot) electrical circuits is prohibited until all power is shut off and grounds are attached.
- All extension cords have grounding prongs.
- Protect flexible cords and cables from damage.
- Use extension cord sets that are three-wire type and imprinted: S, ST, SO, STO.
- All electrical tools and equipment are maintained in safe condition and checked regularly for defects.
- Do not bypass any protective system or device.
- Overhead power lines are located and identified.
- Ensure that ladders never come within 10' (3 m) of electrical power lines.
- Assume that all overhead wires are energized at lethal voltages. Never assume a wire is safe to touch if it is down or appears insulated.
- Never touch a fallen overhead power line.
- Stay at least 10' (3 m) away from overhead wires during cleanup and other activities.
- If an overhead wire falls across your vehicle while driving, stay inside and continue to drive away from the line. If the engine stalls, do not leave vehicle. Call or ask someone to call emergency services.
- Never operate electrical equipment while you are standing in water.
- If working in damp locations, inspect electric cords and equipment and use a ground-fault circuit interrupter.
- Always use caution when working near electricity.

LOCKOUT/TAGOUT PROCEDURES

SHUTDOWN AND LOCKOUT/TAGOUT

1. Notify all affected personnel that work to be performed requires lockout/tagout. Only authorized personnel are permitted to lockout equipment.
2. Prepare for shutdown by reviewing company procedures and identifying all energy sources. Locks and tags shall be uniform and easily identified and only used to lockout equipment for personal protection.
3. Shut down equipment by normal means and isolate the equipment from all energy sources. No one other than the person who installed the lockout tag-out device shall attempt to operate any switch, valve or isolating device bearing a lock or lockout tag.
4. Lockout and tag all energy disconnects or isolating devices. Use individual locks. Primary sources of energy include electricity and pressurized gases or fluids.
5. Release all sources of static or stored energy including mechanical motion, springs, capacitors, thermal energy, and residual pressure in gas or fluid systems.
6. Verify the isolation by operating the normal operating controls.

LOCK REMOVAL AND RETURN TO SERVICE

1. Check that all tools have been removed, all equipment components are intact and that all employees are completely clear of the equipment.
2. Verify that all controls are in a neutral position as to prevent any type of mechanical engagement.
3. Remove all lockout devices and re-energize the equipment. A device may only be removed by the authorized employee who applied it.
4. Notify all affected personnel that the work has been completed and the equipment is back in operation.

WORK AREA CONTROL

- A barrier tape, cones, fence or other methods may be used to surround any work area where a hazard exists that could be accessed by personnel unfamiliar with that work, such as on-lookers.
- The tape shall be prominently placed and completely surround the area where the hazard exists.
- The taped area shall be large enough to permit all persons working within the work area adequate clearance from any hazard.
- Remove all barriers when the hazard is eliminated.
- Metal barrier tape shall not be used.
- There may be a need for attendants to provide additional safety for personnel protection. This may include the use of personnel with flags, detour signs, cones, etc.

OSHA SAFETY COLOR CODES

Color	Examples
Red	Fire protection equipment and apparatus; portable containers of flammable liquids; emergency stop pushbuttons/switches
Yellow	Caution and for marking physical hazards; waste containers for explosive or combustible materials; caution against starting, using or moving equipment under repair; identifies the starting point or power source of machinery
Orange	Dangerous parts of machines; safety starter buttons; the exposed parts of pulleys, gears, rollers, cutting devices and power jaws
Purple	Radiation hazards
Green	Safety areas and location of first aid equipment

TOP OSHA ELECTRICAL VIOLATIONS (2001)			
Section		**Description**	**Total Violations**
1	.404(b)(1)	Electrical—Ground-fault protection	758
2	.405(a)(2)	Electrical—Wiring methods—Temporary wiring	495
3	.404(f)(6)	Electrical—Wiring design and protection—Grounding path	346
4	.405(g)(2)	Electrical—Identification, splices and terminations	304
5	.403(b)(2)	Electrical—Examination, installation and use of equipment	223
6	.403(i)(2)	Electrical—Guarding electrical equipment live parts	174
7	.416(e)(1)	Electrical—Safety-related work practices—Cords and cables	162
8	.405(b)(2)	Electrical—Wiring methods, components and equipment	139
9	.405(b)(1)	Electrical—Wiring methods—Cabinets, boxes and fittings	128
10	.403(b)(1)	Electrical—Approved equipment	117
11	.416(a)(1)	Electrical—Protection of employees	80
12	.405(j)(1)	Electrical—Lighting fixtures, lamps and receptacles	78
13	.405(g)(1)	Electrical—Use of flexible cords and cables	69

Branch Circuits – Ground-Fault Protection

(i) General. The employer shall use either ground-fault circuit interrupters as specified in paragraph (b)(1)(ii) of this section or an assured equipment grounding conductor program as specified in paragraph (b)(1)(iii) of this section to protect employees on construction sites. These requirements are in addition to any other requirements for equipment grounding conductors.

(ii) Ground-fault circuit interrupters. All 120-volt, single-phase, 15- and 20-ampere receptacle outlets on construction sites, which are not a part of the permanent wiring of the building or structure and are in use by employees, shall have approved ground-fault circuit interrupters for personnel protection. Receptacles on a two-wire, single-phase portable or vehicle-mounted generator rated not more than 5 kW, where the circuit conductors of the generator are insulated from the generator frame and all other grounded surfaces, need not be protected with ground-fault circuit interrupters.

(iii) Assured equipment-grounding conductor program. The employer shall establish and implement an assured equipment-grounding conductor program on construction sites covering all cord sets, receptacles that are not a part of the building or structure and equipment connected by cord and plug that are available for use or used by employees. This program shall comply with the following minimum requirements:

(A) A written description of the program, including the specific procedures adopted by the employer, shall be available at the job site for inspection and copying by the assistant secretary and any affected employee.

(B) The employer shall designate one or more competent persons to implement the program.

(C) Each cord set, attachment cap, plug and receptacle of cord sets, and any equipment connected by cord and plug, except cord sets and receptacles that are fixed and not exposed to damage, shall be visually inspected before each day's use for external defects, such as deformed or missing pins or insulation damage, and for indications of possible internal damage. Equipment found to be damaged or defective shall not be used until repaired.

(D) The following tests shall be performed on all cord sets, receptacles that are not a part of the permanent wiring of the building or structure and cord- and plug-connected equipment required to be grounded:

(1) All equipment-grounding conductors shall be tested for continuity and shall be electrically continuous.

(2) Each receptacle and attachment cap or plug shall be tested for correct attachment of the equipment grounding conductor. The equipment-grounding conductor shall be connected to its proper terminal.

(E) All required tests shall be performed:

(1) Before first use;

(2) Before equipment is returned to service following any repairs;

(3) Before equipment is used after any incident that can be reasonably suspected to have caused damage (e.g., when a cord set is run over); and

(4) At intervals not to exceed 3 months, except that cord sets and receptacles that are fixed and not exposed to damage shall be tested at intervals not exceeding 6 months.

(F) The employer shall not make available or permit the use by employees of any equipment that has not met the requirements of paragraph (iii) of this section.

(G) Tests performed as required in this paragraph shall be recorded. This test record shall identify each receptacle, cord set, and cord- and plug-connected equipment that passed the test and shall indicate the last date on which it was tested or the interval for which it was tested. This record shall be kept by means of logs, color coding or other effective means and shall be maintained until replaced by a more current record. The record shall be made available on the job site for inspection by the assistant secretary and any affected employee.

REQUIRED WORKING CLEARANCE — MINIMUM DEPTH IN FEET

Nominal Voltage to Ground	Working Conditions 1	2	3
0 to 150 V	3'	3'	3'
151 to 600 V	3'	4'	4'
601 to 2,500 V	3'	4'	5'
2501 to 9000 V	4'	5'	6'
9001 to 25,000 V	5'	6'	9'
25,001 to 75,000 V	6'	8'	10'
Over 75,000 V	8'	10'	12'

Working Conditions:

Condition 1 – Exposed live parts on one side and no live or grounded parts on the other side. Exposed live parts on both sides effectively guarded by suitable insulating materials.

Condition 2 – Exposed live parts on one side only. Grounded parts on the other side. Concrete, brick, or tile walls are considered grounded.

Condition 3 – Exposed live parts on both sides (not guarded).

MINIMUM ELEVATION OF UNGUARDED LIVE PARTS ABOVE WORKING SPACE IN FEET

Nominal Voltage Between Phases	Elevation
601 to 7,500 V	9'
7,501 to 35,000 V	9½'
Over 35,000 V	9½' + 0.37" for every 1,000 volts over 35,000 volts

MINIMUM CLEARANCE — LIVE PARTS

*Minimum Clearance of Live Parts (inches)				Nominal Voltage Rating (kV)	Impulse Withstand B.I.L. (kV)	
Phase-To-Phase		Phase-To-Ground				
Indoors	Outdoors	Indoors	Outdoors		Indoors	Outdoors
4.5	7	3.0	6	2.4–4.16	60	95
5.5	7	4.0	6	7.2	75	95
7.5	12	5.0	7	13.8	95	110
9.0	12	6.5	7	14.4	110	110
10.5	15	7.5	10	23	125	150
12.5	15	9.5	10	34.5	150	150
18.0	18	13.0	13	—	200	200
—	18	—	13	46	—	200
—	21	—	17	—	—	250
—	21	—	17	69	—	250
—	31	—	25	—	—	350
—	53	—	42	115	—	550
—	53	—	42	138	—	550
—	63	—	50	—	—	650
—	63	—	50	161	—	650
—	72	—	58	—	—	750
—	72	—	58	230	—	750
—	89	—	71	—	—	900
—	105	—	83	—	—	1050

*This represents the minimum clearance for rigid parts and bare conductors under favorable conditions. They shall be increased for conductor movement, unfavorable conditions or where space is limited. Impulse withstand voltage for a particular system voltage is determined by the type of surge protection equipment that is utilized.

PORTABLE LADDER SAFETY

- Read and follow all labels and markings.
- Avoid electrical hazards! Look for overhead power lines before handling a ladder. Never use a metal ladder near power lines.
- Always inspect the ladder before using it.
- Always maintain a three point (two hands and a foot, or two feet and a hand) contact on the ladder when climbing.
- Only use ladders and appropriate accessories for their designated purposes.
- Ladders must be free of any slippery material on the rungs, steps or feet.
- Do not use a self-supporting ladder as a single ladder or in a partially closed position.
- Do not use the top step/rung of a ladder.
- Use a ladder only on a stable and level surface.
- Never place a ladder on a box, barrel or other unstable base to obtain additional height.
- Never move or shift a ladder while a person or equipment is on it.
- An extension or straight ladder used to access an elevated surface must extend at least 3' above the point of support.
- Never stand on the three top rungs of a straight, single or extension ladder.
- The proper angle for setting up a ladder is to place its base a quarter of the working length of the ladder from the wall or other vertical surface.

PORTABLE LADDER SAFETY *(cont.)*

- A ladder placed in any location where it can be displaced by other work activities must be secured, or a barricade must be erected.
- Be sure that all locks are properly engaged.
- Never exceed the maximum load rating of a ladder. Be aware of the weight it is supporting (including tools).

ELECTRICAL GLOVE INSPECTION

- Gloves shall be inspected before and after every use.
- Check that the voltage rating of the insulating equipment is greater than the application voltage.
- Check for defects by stretching a small area at a time.
- Look for cracks, cuts, scratches, embedded material or weak spots.
- Inspect the entire glove. Any detectable flaw is cause for gloves to be returned for testing.
- Check for pinhole leaks by trapping air in the fingers of the glove. Also squeeze the cuff with one hand and inspect the palm, thumb and fingers for defects.
- Hold the glove to your face to check for air leakage.
- Any leakage is cause for the glove to be destroyed. Cut the glove in half immediately.
- Check the test date stamped (month and year) on any rubber protective equipment.
- Maximum time between tests shall be 6 months for gloves and 12 months for sleeves and insulating blankets.

TREE TRIMMING AND REMOVAL SAFETY

Assume All Power Lines Are Energized!

- Contact the utility company to discuss de-energizing and grounding or shielding of power lines.
- All tree trimming and removal work within 10' of a power line must be done by trained and experienced line-clearance tree trimmers. A second tree trimmer is required within normal voice communication range.
- Line-clearance tree trimmers must be aware of and maintain the proper minimum approach distances when working around energized power lines.
- Use extreme caution when moving ladders and equipment around downed trees and power lines.

Stay Alert at All Times!

- Do not trim trees in dangerous weather conditions.
- Perform a hazard assessment of the work area before starting work.
- Eliminate or minimize exposure to hazards at the tree and in the surrounding area.
- Operators of chain saws and other equipment should be trained and the equipment properly maintained.
- Use personal protective equipment such as gloves, safety glasses, hard hats, hearing protection, etc., as recommended in the equipment manufacturers' operating manuals.

PORTABLE GENERATOR SAFETY

Major Causes of Injuries and Fatalities

- Shocks and electrocution from improper use or accidentally energizing other equipment.
- Carbon monoxide from a generator's exhaust.
- Fires from improperly refueling the generator.

Safe Work Practices

- Inspect portable generators for damage or loose fuel lines.
- Keep the generator dry.
- Maintain and operate portable generators in accordance with the manufacturer's manual.
- Never attach a generator directly to the electrical system of a structure unless the generator has a properly installed transfer switch.
- Always plug electrical appliances directly into the generator using the manufacturer's supplied cords. Use undamaged heavy-duty extension cords that are grounded (three-pronged).
- Use ground-fault circuit interrupters.
- Before refueling, shut down the generator. Never store fuel indoors.

Carbon Monoxide Poisoning

Many people have died from CO poisoning because their generator was not adequately ventilated.

- Never use a generator indoors.
- Never place a generator outdoors near doors, windows or vents.
- If you or others show these symptoms—dizziness, headaches, nausea, tiredness—get to fresh air immediately and seek medical attention.

FIRST AID GUIDE

FIRST — CHECK THE SCENE AND THE INJURED PERSON.
THEN — CALL 911 OR THE LOCAL EMERGENCY NUMBER.
THEN — CARE FOR THE INJURED PERSON.

BLEEDING

1) Direct Pressure and Elevation —
   - Place dressing and apply pressure directly over wound, then elevate above the level of the heart unless there is evidence of a fracture.
2) Apply pressure bandage —
   - Wrap bandage tightly over the dressing.
3) Pressure Points —
   - If bleeding doesn't stop after direct pressure, elevation and the pressure bandage, compress the pressure point.
   - Arm — Push the brachial artery against the upper arm bone.
   - Leg — Apply pressure on femoral artery, pushing it against the femur or pelvic bone.
4) Nosebleed —
   - Be sure to have the victim lean forward and pinch the nostrils together until bleeding stops.

POISONING

Signs: Vomiting, heavy labored breathing, sudden onset of pain or illness, burns or odor around the lips or mouth, extreme or bizarre behavior.

- **Call the poison control center or local emergency number and follow their directions.**
- Try to identify the poison and inform the poison control center of the type, when incident occurred, victim's age, size and symptoms, and quantity of poison that may have been ingested, inhaled, absorbed, or injected.

If unconscious or nauseous

- Position victim on side and monitor vital signs.
- **DO NOT** give anything by mouth.

FIRST AID GUIDE *(cont.)*

SHOCK

Signs: Cool, moist, pale, bluish skin, weak rapid pulse (100+), increased breathing rate, lethargic, nausea.

SHOCK CARE

1) Have the victim lie down in a comfortable position to minimize pain, which can intensify the body's stress and accelerate the progression of shock.
2) Control any external bleeding
3) Maintain normal body temperature.
4) Elevate legs a foot unless there are possible head, neck, back injuries or broken bones involving the hips or legs. If unsure, leave the victim lying flat.
5) Do not give anything by mouth.
6) Call 911 or your local emergency number at once.

BURNS

Signs: For thin burns: Redness, pain and swelling.
For deep burns: blisters, deep tissue destruction, and charred appearance.

1) Put out the heat — Remove the victim from the source of the burn.
2) Cool all burns (except electrical) — Run or pour cool water on burn. Immerse if possible.
3) Cover the burn — Use dry, sterile dressing and bandage.
4) Keep victim comfortable — Not chilled or over heated.

Chemical burns: must be flushed with large amounts of water until help arrives.

Electrical burn: Be absolutely sure power is **OFF** before touching the victim.

ELECTRICAL SHOCK

Signs: Unconsciousness, absence of breathing and/or pulse.

1) **TURN OFF THE POWER SOURCE AND CALL 911.**
2) **DO NOT** approach the victim until power de energized.
3) **DO NOT** move victim unless necessary.
4) Administer **CPR** if necessary.
5) Treat for shock and check for other injuries.

FROSTBITE

Signs: Flushed, white or gray skin. Pain may be felt early, then subside. Blisters may appear.

1) Cover the frozen part. Loosen clothing or foot wear.
2) Give the victim a warm drink. (No alcohol or caffeine.)
3) Immerse frozen part in warm water up to 105°, or wrap in warm blankets.
4) Remove from water and discontinue once part becomes flushed.
5) Elevate the area.
6) **DO NOT** rub the frozen part. **DO NOT** break the blisters. **DO NOT** use extreme or dry heat to rewarm the part.
7) If fingers or toes involved, place gauze between them when bandaging.

HYPOTHERMIA

Signs: Lowered body core temperature. Persistent shivering, lips blue, slow slurred speech, memory lapses.

1) Move victim to shelter and remove any wet clothing.
2) Rewarm victim with blankets or body-to-body contact.
3) Give warm liquids if possible and keep victim quiet.
4) **DO NOT** give alcohol or caffeine.
5) Monitor victim and give **CPR** if necessary.

HEAT EXHAUSTION/HEAT STROKE

Signs: *Heat Exhaustion:* Pale, clammy skin, profuse perspiration, weakness, nausea, headache.

Heat Stroke. Hot dry red skin, no perspiration, rapid pulse. High body temperature (105°+).

1) Call 911.
2) Get the victim out of the heat.
3) Loosen tight or restrictive clothing.
4) Remove perspiration soaked clothing.
5) Apply cool, wet cloths to the skin and fan victim.
6) Give cool water if possible.

FIRST AID GUIDE *(cont.)*

NOT BREATHING

1. CHECK THE VICTIM.

If no response . . .

2. CALL 911.

3. CARE FOR THE VICTIM.

Look, listen and feel for breathing for 5 to 7 seconds.

If the person is not breathing

Step 1: Position victim on back and support head and neck.
Step 2: Tilt head back and lift chin.
Step 3: Look, listen, and feel for breathing for 5 seconds.

STILL NOT BREATHING

Step 1: Give two slow gentle breaths.
Step 2: Check pulse for 5 to 7 seconds.
Step 3: Check for profuse bleeding.

4. BEGIN RESCUE BREATHING.

If pulse is present but person is still not breathing . . .

Step 1: Give one slow breath every five seconds for one minute.
Step 2: Recheck pulse and breathing every minute.
Step 3: **Continue rescue breathing as long as pulse is present but person is not breathing.**

5. BEGIN CPR IF NO SIGN OF PULSE AND NO BREATHING.

CHOKING

1. Check the victim.

Step 1: Ask, **"Are you choking?"** If victim cannot respond, breathe, is coughing weakly or making noises . . .
Step 2: **Phone 911.**
Step 3: **Do Abdominal Thrusts.**

A. Wrap your arms around victims's waist. Make a first. Place thumbside of fist against middle of abdomen just above the navel. Grasp fist with other hand.
B. Give quick, upward thrusts.

Repeat until object is dislodged or person becomes unconscious.

TYPES OF FIRE EXTINGUISHERS

TYPE A: To extinguish fires involving trash, cloth, paper, and other wood- or pulp-based materials. The flames are put out by water-based ingredients or dry chemicals.

TYPE B: To extinguish fires involving greases, paints, solvents, gas, and other petroleum-based liquids. The flames are put out by cutting off oxygen and stopping the release of flammable vapors. Dry chemicals, foams, and halon are used.

TYPE C: To extinguish fires involving electricity. The combustion is put out the same way as with a type B extinguisher but, most importantly, the chemical in a type C MUST be nonconductive to electricity in order to be safe and effective.

TYPE D: To extinguish fires involving combustible metals. Please be advised to obtain important information from your local fire department on the requirements for type D fire extinguishers for your area.

Any combination of letters indicate that an extinguisher will put out more than one type of fire. Type BC will put out two types of fires. The size of the fire to be extinguished is shown by a number in front of the letter such as 100A. For example:

Class 1A will extinguish 25 burning sticks 40 inches long.

Class 1B will extinguish a paint thinner fire 2.5 square feet in size.

Class 100B will put out a fire 100 times larger than type 1B.

Here are some basic guidelines to follow:

- By using a type ABC you will cover most basic fires.
- Use fire extinguishers with a gauge and that are constructed with metal. Also note if the unit is U.L. approved.
- Utilize more than one extinguisher and be sure that each unit is mounted in a clearly visible and accessible manner.
- After purchasing any fire extinguisher always review the basic instructions for its intended use. Never deviate from the manufacturer's guidelines. Following this simple procedure could end up saving lives.

CHAPTER 10
Materials and Tools

STRENGTH GAIN VS. PULL ANGLE

The weight bearing capacity of a strap, for example, increases by the factor K shown below as the angle of the strap decreases. A 100 lb. capacity strap at a 60° pulling angle loses 50 lbs. of weight bearing ability. At perfectly vertical, the ability is 100%.

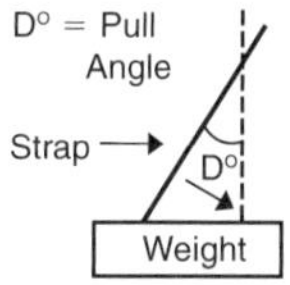

K	D°	K	D°	K	D°
.7412	75	.3572	50	.0937	25
.6580	70	.2929	45	.0603	20
.5774	65	.2340	40	.0341	15
.5000	60	.1340	35	.0152	10
.4264	55	.1208	30	.0038	5

LENGTH OF WIRE CABLE PER REEL

Cable length (feet) = X (X + Y) Z (K)
Use equation above to determine the length of wire cable that is wound smoothly on a reel. The dimensions for X, Y and Z are in inches.

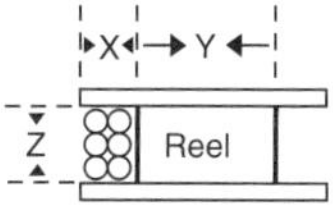

(K)	Cable Dia. (Inches)	(K)	Cable Dia. (Inches)
.0476	2.250	.239	1.000
.0532	2.125	.308	.875
.0597	2.000	.428	.750
.0675	1.875	.607	.625
.0770	1.750	.741	.563
.0886	1.625	.925	.500
.107	1.500	1.19	.438
.127	1.375	1.58	.375
.152	1.250	2.21	.313
.191	1.125	3.29	.250

SHEET METAL SCREW CHARACTERISTICS				
Screw Size #	**Screw Dia. (Inches)**	**Diameter of Pierced Hole (Inches)**	**Hole Size #**	**Thickness of Metal – Gauge #**
4	.112	.086	44	28
		.086	44	26
		.093	42	24
		.098	42	22
		.100	40	20
6	.138	.111	39	28
		.111	39	26
		.111	39	24
		.111	38	22
		.111	36	20
7	.155	.121	37	28
		.121	37	26
		.121	35	24
		.121	33	22
		.121	32	20
		—	31	18
8	.165	.137	33	26
		.137	33	24
		.137	32	22
		.137	31	20
		—	30	18
10	.191	.158	30	26
		.158	30	24
		.158	30	22
		.158	29	20
		.158	25	18
12	.218	—	26	24
		.185	25	22
		.185	24	20
		.185	22	18
14	.251	—	15	24
		.212	12	22
		.212	11	20
		.212	9	18
Deviations in materials and conditions could require variations from these dimensions.				

WOOD SCREW CHARACTERISTICS

Screw Size #	Wood Screw Standard Lengths (Inches)	Size of Pilot Hole		Size of Shank Hole	
		Softwood Bit #	Hardwood Bit #	Clearance Bit #	Hole Diameter (Inches)
0	1/4	75	66	52	.060
1	1/4 to 3/8	71	57	47	.073
2	1/4 to 1/2	65	54	42	.086
3	1/4 to 5/8	58	53	37	.099
4	3/8 to 3/4	55	51	32	.112
5	3/8 to 3/4	53	47	30	.125
6	3/8 to 1 1/2	52	44	27	.138
7	3/8 to 1 1/2	51	39	22	.151
8	1/2 to 2	48	35	18	.164
9	5/8 to 2 1/4	45	33	14	.177
10	5/8 to 2 1/2	43	31	10	.190
11	3/4 to 3	40	29	4	.203
12	7/8 to 3 1/2	38	25	2	.216
14	1 to 4 1/2	32	14	D	.242
16	1 1/4 to 5 1/2	29	10	I	.268
18	1 1/2 to 6	26	6	N	.294
20	1 3/4 to 6	19	3	P	.320
24	3 1/2 to 6	15	D	V	.372

HEX HEAD BOLT AND TORQUE CHARACTERISTICS				
BOLT MAKE-UP IS STEEL WITH COARSE THREADS				
Number of Threads Per Inch	**Hex Head Bolt Size (Inches)**	**Torque = Foot-Pounds**		
		 **SAE 0-1-2 74,000 psi**	 **SAE Grade 3 100,000 psi**	 **SAE Grade 5 120,000 psi**
4.5	2	2750	5427	4550
5	1 3/4	1900	3436	3150
6	1 1/2	1100	1943	1775
6	1 3/8	900	1624	1500
7	1 1/4	675	1211	1105
7	1 1/8	480	872	794
8	1	310	551	587
9	7/8	206	372	382
10	3/4	155	234	257
11	5/8	96	145	154
12	9/16	69	103	114
13	1/2	47	69	78
14	7/16	32	47	54
16	3/8	20	30	33
18	5/16	12	17	19
20	1/4	6	9	10
For fine thread bolts, increase by 9%.				

HEX HEAD BOLT AND TORQUE CHARACTERISTICS (*cont.*)

BOLT MAKE-UP IS STEEL WITH COARSE THREADS				
Number of Threads Per Inch	Hex Head Bolt Size (Inches)	Torque = Foot-Pounds		
		SAE Grade 6 133,000 psi	SAE Grade 7 133,000 psi	SAE Grade 8 150,000 psi
4.5	2	7491	7500	8200
5	1¾	5189	5300	5650
6	1½	2913	3000	3200
6	1⅜	2434	2500	2650
7	1¼	1815	1825	1975
7	1⅛	1304	1325	1430
8	1	825	840	700
9	⅞	550	570	600
10	¾	350	360	380
11	⅝	209	215	230
12	9/16	150	154	169
13	½	106	110	119
14	7/16	69	71	78
16	⅜	43	44	47
18	5/16	24	25	29
20	¼	12.5	13	14

For fine thread bolts, increase by 9%.
For special alloy bolts, obtain torque rating from the manufacturer.

WHITWORTH HEX HEAD BOLT AND TORQUE CHARACTERISTICS

BOLT MAKE-UP IS STEEL WITH COARSE THREADS

Number of Threads Per Inch	Whitworth Type Hex Head Bolt Size (Inches)	Torque = Foot-Pounds			
		Grades A & B 62,720 psi	Grade S 112,000 psi	Grade T 123,200 psi	Grade V 145,600 psi
8	1	276	497	611	693
9	7/8	186	322	407	459
11	3/4	118	213	259	287
11	5/8	73	128	155	175
12	9/16	52	94	111	128
12	1/2	36	64	79	89
14	7/16	24	43	51	58
16	3/8	15	27	31	36
18	5/16	9	15	18	21
20	1/4	5	7	9	10

For fine thread bolts, increase by 9%.

METRIC HEX HEAD BOLT AND TORQUE CHARACTERISTICS

BOLT MAKE-UP IS STEEL WITH COARSE THREADS

Thread Pitch (millimeters)	Bolt Size (millimeters)	Torque = Foot-Pounds			
		5D Standard 5D 71,160 psi	8G Standard 8G 113,800 psi	10K Standard 10K 142,000 psi	12K Standard 12K 170,674 psi
3.0	24	261	419	570	689
2.5	22	182	284	394	464
2.0	18	111	182	236	183
2.0	16	83	132	175	208
1.25	14	55	89	117	137
1.25	12	34	54	70	86
1.25	10	19	31	40	49
1.0	8	10	16	22	27
1.0	6	5	6	8	10

For fine thread bolts, increase by 9%.

ALLEN HEAD AND MACHINE SCREW BOLT AND TORQUE CHARACTERISTICS

Number of Threads Per Inch	Allen Head and Mach. Screw Bolt Size	Torque in Foot-Pounds or Inch-Pounds		
		Allen Head Case H Steel 160,000 psi	Mach. Screw Yellow Brass 60,000 psi	Mach. Screw Silicone Bronze 70,000 psi
4.5	2"	8800	–	–
5	$1\frac{3}{4}$"	6100	–	–
6	$1\frac{1}{2}$"	3450	655	595
6	$1\frac{3}{8}$"	2850	–	–
7	$1\frac{1}{4}$"	2130	450	400
7	$1\frac{1}{8}$"	1520	365	325
8	1"	970	250	215
9	$\frac{7}{8}$"	640	180	160
10	$\frac{3}{4}$"	400	117	104
11	$\frac{5}{8}$"	250	88	78
12	$\frac{9}{16}$"	180	53	49
13	$\frac{1}{2}$"	125	41	37
14	$\frac{7}{16}$"	84	30	27
16	$\frac{3}{8}$"	54	20	17
18	$\frac{5}{16}$"	33	125 in#	110 in#
20	$\frac{1}{4}$"	16	70 in#	65 in#
24	#10	60	22 in#	20 in#
32	#8	46	19 in#	16 in#
32	#6	21	10 in#	8 in#
40	#5	—	7.2 in#	6.4 in#
40	#4	—	4.9 in#	4.4 in#
48	#3	—	3.7 in#	3.3 in#
56	#2	—	2.3 in#	2 in#

For fine thread bolts, increase by 9%.

TIGHTENING TORQUE IN POUND-FEET-SCREW FIT			
Wire Size, AWG/kcmil	**Driver**	**Bolt**	**Other**
18–16	1.67	6.25	4.2
14–8	1.67	6.25	6.125
6–4	3.0	12.5	8.0
3–1	3.2	21.00	10.40
0–2/0	4.22	29	12.5
3/0–200	–	37.5	17.0
250–300	–	50.0	21.0
400	–	62.5	21.0
500	–	62.5	25.0
600–750	–	75.0	25.0
800–1000	–	83.25	33.0
1250–2000	–	83.26	42.0

SCREW TORQUES	
Screw Size in Inches Across, Hex Flats	**Torque in Pound-Feet**
1/8	4.2
5/32	8.3
3/16	15
7/32	23.25
1/4	42

STANDARD TAPS AND DIES

Thread Size	Coarse			Fine		
	Drill Size	Threads Per Inch	Decimal Size	Drill Size	Threads Per Inch	Decimal Size
4"	3	4	3.75"	–	–	–
3¾"	3	4	3.5"	–	–	–
3½"	3	4	3.25"	–	–	–
3¼"	3	4	3.0"	–	–	–
3"	2	4	2.75"	–	–	–
2¾"	2	4	2.5"	–	–	–
2½"	2	4	2.25"	–	–	–
2¼"	2	4.5	2.0313"	–	–	–
2"	1	4.5	1.7813"	–	–	–
1¾"	1	2	1.5469"	–	–	–
1½"	1	6	1.3281"	1 27/64"	12	1.4219"
1⅜"	1	6	1.2188"	1 19/64"	12	1.2969"
1¼"	1	7	1.1094"	1 11/64"	12	1.1719"
1⅛"	63/64"	7	.9844"	1 3/64"	12	1.0469"
1"	⅞"	8	.8750"	15/16"	14	.9375"
⅞"	49/64"	9	.7656"	13/16"	14	.8125"
¾"	21/32"	10	.6563"	11/16"	16	.6875"
⅝"	17/32"	11	.5313"	37/64"	18	.5781"
9/16"	31/64"	12	.4844"	33/64"	18	.5156"
½"	27/64"	13	.4219"	29/64"	20	.4531"
7/16"	U	14	.368"	25/64"	20	.3906"
⅜"	5/16"	16	.3125"	Q	24	.332"
5/16"	F	18	.2570"	I	24	.272"
¼"	#7	20	.201"	#3	28	.213"
#12	#16	24	.177"	#14	28	.182"
#10	#25	24	.1495"	#21	32	.159"
3/16"	#26	24	.147"	#22	32	.157"
#8	#29	32	.136"	#29	36	.136"
#6	#36	32	.1065"	#33	40	.113"
#5	#38	40	.1015"	#37	44	.104"
⅛"	3/32"	32	.0938"	#38	40	.1015"
#4	#43	40	.089"	#42	48	.0935"
#3	#47	48	.0785"	#45	56	.082"
#2	#50	56	.07"	#50	64	.07"
#1	#53	64	.0595"	#53	72	.0595"
#0	–	–	–	3/64"	80	.0469"

TAPS AND DIES — METRIC CONVERSIONS

Thread Pitch (mm)	Fine Thread Size		Tap Drill Size	
	Inches	mm	Inches	mm
4.5	1.6535	42	1.4567	37.0
4.0	1.5748	40	1.4173	36.0
4.0	1.5354	39	1.3779	35.0
4.0	1.4961	38	1.3386	34.0
4.0	1.4173	36	1.2598	32.0
3.5	1.3386	34	1.2008	30.5
3.5	1.2992	33	1.1614	29.5
3.5	1.2598	32	1.1220	28.5
3.5	1.1811	30	1.0433	26.5
3.0	1.1024	28	.9842	25.0
3.0	1.0630	27	.9449	24.0
3.0	1.0236	26	.9055	23.0
3.0	.9449	24	.8268	21.0
2.5	.8771	22	.7677	19.5
2.5	.7974	20	.6890	17.5
2.5	.7087	18	.6102	15.5
2.0	.6299	16	.5118	14.0
2.0	.5512	14	.4724	12.0
1.75	.4624	12	.4134	10.5
1.50	.4624	12	.4134	10.5
1.50	.3937	11	.3780	9.6
1.50	.3937	10	.3386	8.6
1.25	.3543	9	.3071	7.8
1.25	.3150	8	.2677	6.8
1.0	.2856	7	.2362	6.0
1.0	.2362	6	.1968	5.0
.90	.2165	5.5	.1811	4.6
.80	.1968	5	.1653	4.2
.75	.1772	4.5	.1476	3.75
.70	.1575	4	.1299	3.3
.75	.1575	4	.1279	3.25
.60	.1378	3.5	.1142	2.9
.60	.1181	3	.0945	2.4
.50	.1181	3	.0984	2.5
.45	.1124	2.6	.0827	2.1
.45	.0984	2.5	.0787	2.0
.40	.0895	2.3	.0748	1.9
.40	.0787	2	.0630	1.6
.45	.0787	2	.0590	1.5
.35	.0590	1.5	.0433	1.1

RECOMMENDED DRILLING SPEEDS		
Material	**Bit Sizes**	**RPM Speed Range**
Glass	Special Metal Tube Drilling	700 Only
Plastics	7/16" and larger 3/8" 5/16" 1/4" 3/16" 1/8" 1/16" and smaller	500 – 1000 1500 – 2000 2000 – 2500 3000 – 3500 3500 – 4000 5000 – 6000 6000 – 6500
Woods	1" and larger 3/4" to 1" 1/2" to 3/4" 1/4" to 1/2" 1/4" and smaller carving / routing	700 – 2000 2000 – 2300 2300 – 3100 3100 – 3800 3800 – 4000 4000 – 6000
Soft Metals	7/16" and larger 3/8" 5/16" 1/4" 3/16" 1/8" 1/16" and smaller	1500 – 2500 3000 – 3500 3500 – 4000 4500 – 5000 5000 – 6000 6000 – 6500 6000 – 6500
Steel	7/16" and larger 3/8" 5/16" 1/4" 3/16" 1/8" 1/16" and smaller	500 – 1000 1000 – 1500 1000 – 1500 1500 – 2000 2000 – 2500 3000 – 4000 5000 – 6500
Cast Iron	7/16" and larger 3/8" 5/16" 1/4" 3/16" 1/8" 1/16" and smaller	1000 – 1500 1500 – 2000 1500 – 2000 2000 – 2500 2500 – 3000 3500 – 4500 6000 – 6500

TORQUE LUBRICATION EFFECTS IN FOOT-POUNDS

Lubricant	5/16" - 18 Thread	1/2" - 13 Thread	Torque Decrease
Graphite	13	62	49 – 55%
Mily Film	14	66	45 – 52%
White Grease	16	79	35 – 45%
Sae 30	16	79	35 – 45%
Sae 40	17	83	31 – 41%
Sae 20	18	87	28 – 38%
Plated	19	90	26 – 34%
No Lube	29	121	0%

METALWORKING LUBRICANTS

Materials	Threading	Lathing	Drilling
Machine Steels	Dissolvable Oil Mineral Oil Lard Oil	Dissolvable Oil	Dissolvable Oil Sulpherized Oil Min. Lard Oil
Tool Steels	Lard Oil Sulpherized Oil	Dissolvable Oil	Dissolvable Oil Sulpherized Oil
Cast Irons	Sulpherized Oil Dry Min. Lard Oil	Dissolvable Oil Dry	Dissolvable Oil Dry Air Jet
Malleable Irons	Soda Water Lard Oil	Soda Water Dissolvable Oil	Soda Water Dry
Aluminums	Kerosene Dissolvable Oil Lard Oil	Dissolvable Oil	Kerosene Dissolvable Oil
Brasses	Dissolvable Oil Lard Oil	Dissolvable Oil	Kerosene Dissolvable Oil Dry
Bronzes	Dissolvable Oil Lard Oil	Dissolvable Oil	Dissolvable Oil Dry
Coppers	Dissolvable Oil Lard Oil	Dissolvable Oil	Kerosene Dissolvable Oil Dry

TYPES OF SOLDERING FLUX

To Solder	Use
Cast iron	Cuprous oxide
Galvanized iron, galvanized, steel	Hydrochloric acid
Brass, copper, gold, iron, silver, steel	Borax
Bronze, cadmium, copper, lead	Resin
Brass, copper, gun metal, nickel, tin, zinc	Ammonia chloride
Bismuth, brass, copper, gold, silver, tin	Zinc chloride
Silver	Sterling
Pewter and Lead	Tallow
Stainless	Stainless steel (only)

HARD SOLDER ALLOYS

To hard solder	Copper %	Gold %	Silver %	Zinc %
Gold	22	67	11	–
Silver	20	–	70	10
Hard brass	45	–	–	55
Soft brass	22	–	–	78
Copper	50	–	–	50
Cast iron	55	–	–	45
Steel and iron	64	–	–	36

SOFT SOLDER ALLOYS

To soft solder	Lead %	Tin %	Zinc %	Bism %	Other %
Gold	33	67	–	–	–
Silver	33	67	–	–	–
Brass	34	66	–	–	–
Copper	40	60	–	–	–
Steel and iron	50	50	–	–	–
Galvanized steel	42	58	–	–	–
Tinned steel	36	64	–	–	–
Zinc	45	55	–	–	–
Block Tin	1	99	–	–	–
Lead	67	33	–	–	–
Gun metal	37	63	–	–	–
Pewter	25	25	–	50	–
Bismuth	33	33	–	34	–
Aluminum	–	70	25	–	5

PROPERTIES OF WELDING GASES

Type of Gas	Characteristics	Common Tank Sizes (cu. ft.)
Acetylene	C_2H_2, explosive gas, flammable, garlic - like odor, colorless, dangerous if used in pressures over 15 psig (30 psig absolute)	10, 40, 75 100, 300
Argon	Ar, non-explosive inert gas, tasteless, odorless, colorless	131, 330 4754 (Liquid)
Carbon Dioxide	CO_2, non-explosive inert gas, tasteless, odorless, colorless (in large quantities is toxic)	20 lbs., 50 lbs.
Helium	He, non-explosive inert gas, tasteless, odorless, colorless	221
Hydrogen	H_2, explosive gas, tasteless, odorless, colorless	191
Nitrogen	N_2, non-explosive inert gas, tasteless, odorless, colorless	20, 40, 80 113, 225
Oxygen	O_2, non-explosive gas, tasteless, odorless, colorless, supports combustion	20, 40, 80 122, 244 4500 (liquid)

WELDING RODS – 36" LONG

Rod Size (Dia. In.)	Number of Rods Per Pound			
	Aluminum	Brass	Cast Iron	Steel
3⁄8"	–	1.0	.25	1.0
5⁄16"	–	–	.50	1.33
1⁄4"	6.0	2.0	2.25	2.0
3⁄16"	9.0	3.0	5.50	3.5
5⁄32"	–	–	–	5.0
1⁄8"	23.0	7.0	–	8.0
3⁄32"	41.0	13.0	–	14.0
1⁄16"	91.0	29.0	–	31.0

PULLEY AND GEAR FORMULAS

For single reduction or increase of speed by means of belting where the speed at which each shaft should run is known, and one pulley is in place:

Multiply the diameter of the pulley which you have by the number of revolutions per minute that its shaft makes; divide this product by the speed in revolutions per minute at which the second shaft should run. The result is the diameter of pulley to use.

Where both shafts with pulleys are in operation and the speed of one is known:

Multiply the speed of the shaft by diameter of its pulley and divide this product by diameter of pulley on the other shaft. The result is the speed of the second shaft.

Where a countershaft is used to obtain size of main driving or driven pulley, or speed of main driving or driven shaft, it is necessary to calculate, as above, between the known end of the transmission and the countershaft, then repeat this calculation between the countershaft and the unknown end.

A set of gears of the same pitch transmits speeds in proportion to the number of teeth they contain. Count the number of teeth in the gear wheel and use this quantity instead of the diameter of pulley, mentioned above, to obtain number of teeth cut in unknown gear, or speed of second shaft.

Formulas for Finding Pulley Sizes:

$$d = \frac{D \times S}{s'} \qquad D = \frac{d \times s'}{S}$$

d = diameter of driven pulley

D = diameter of driving pulley

s' = number of revolutions per minute of driven pulley

S = number of revolutions per minute of driving pulley

PULLEY AND GEAR FORMULAS *(cont.)*

Formulas for Finding Gear Sizes:

$$n = \frac{N \times S}{s'} \qquad N = \frac{n \times s'}{S}$$

n = number of teeth in pinion (driving gear)

N = number of teeth in gear (driven gear)

s' = number of revolutions per minute of gear

S = number of revolutions per minute of pinion

Formula to Determine Shaft Diameter:

$$\text{diameter of shaft in inches} = \sqrt[3]{\frac{K \times HP}{RPM}}$$

HP = the horsepower to be transmitted

RPM = speed of shaft

K = factor which varies from 50 to 125 depending on type of shaft and distance between supporting bearings.

For line shaft having bearings 8 feet apart:

K = 90 for turned shafting

K = 70 for cold-rolled shafting

Formula to Determine Belt Length:

$$\text{length of belt} = \frac{3.14\,(D + d)}{2} + 2\left(\sqrt{X^2 + \left(\frac{D - d}{2}\right)^2}\right)$$

D = diameter of large pulley

d = diameter of small pulley

X = distance between centers of shafting

STANDARD V BELT LENGTHS IN INCHES

A BELTS

Standard Belt No.	Pitch Length	Outside Length
A26	27.3	28.0
A31	32.3	33.0
A35	36.3	37.0
A38	39.3	40.0
A42	43.3	44.0
A46	47.3	48.0
A51	52.3	53.0
A55	56.3	57.0
A60	61.3	62.0
A68	69.3	70.0
A75	76.3	77.0
A80	81.3	82.0
A85	86.3	87.0
A90	91.3	92.0
A96	97.3	98.0
A105	106.3	107.0
A112	113.3	114.0
A120	121.3	122.0
A128	129.3	130.0

B BELTS

Standard Belt No.	Pitch Length	Outside Length
B35	36.8	38.0
B38	39.8	41.0
B42	43.8	45.0
B46	47.8	49.0
B51	52.8	54.0
B55	56.8	58.0
B60	61.8	63.0
B68	69.8	71.0
B75	76.8	78.0
B81	82.8	84.0
B85	86.8	88.0
B90	91.8	93.0
B97	98.8	100.0
B105	106.8	108.0
B112	113.8	115.0
B120	121.8	123.0
B128	129.8	131.0
B136	137.8	139.0
B144	145.8	147.0
B158	159.8	161.0
B173	174.8	176.0
B180	181.8	183.0
B195	196.8	198.0
B210	211.8	213.0
B240	240.3	241.5
B270	270.3	271.5
B300	300.3	301.5

C BELTS

Standard Belt No.	Pitch Length	Outside Length
C51	53.9	55.0
C60	62.9	64.0
C68	70.9	81.0
C75	77.9	79.0
C81	83.9	85.0
C85	87.9	89.0
C90	92.9	94.0
C96	98.9	100.0
C105	107.9	109.0
C112	114.9	116.0
C120	122.9	124.0
C128	130.9	132.0
C136	138.9	140.0
C144	146.9	148.0
C158	160.9	162.0
C162	164.9	166.0
C173	175.9	177.0
C180	182.9	184.0
C195	197.9	199.0
C210	212.9	214.0
C240	240.9	242.0
C270	270.9	272.0
C300	300.9	302.0
C360	360.9	362.0
C390	390.9	392.0
C420	420.9	422.0

D BELTS

Standard Belt No.	Pitch Length	Outside Length
D120	123.3	125.0
D128	131.3	133.0
D144	147.3	149.0
D158	161.3	163.0
D162	165.3	167.0
D173	176.3	178.0
D180	183.3	185.0
D195	198.3	200.0
D210	213.3	215.0
D240	240.8	242.0
D270	270.8	272.5
D300	300.8	302.5
D330	330.8	332.5
D360	360.8	362.5
D390	390.8	392.5
D420	420.8	422.5
D480	480.8	482.5
D540	540.8	542.5
D600	600.8	602.5

E BELTS

Standard Belt No.	Pitch Length	Outside Length	Standard Belt No.	Pitch Length	Outside Length
E180	184.5	187.5	E360	361.0	364.0
E195	199.5	202.5	E390	391.0	394.0
E210	214.5	217.5	E420	421.0	424.0
E240	241.0	244.0	E480	481.0	484.0
E270	271.0	274.0	E540	541.0	544.0
E300	301.0	304.0	E600	601.0	604.0
E330	331.0	334.0			

STANDARD V BELT LENGTHS IN INCHES *(cont.)*

3 V Belts		5 V Belts		8 V Belts	
3V250	25.0	5V500	50.0	8V1000	100.0
3V265	26.5	5V530	53.0	8V1060	106.0
3V280	28.0	5V560	56.0	8V1120	112.0
3V300	30.0	5V600	60.0	8V1180	118.0
3V315	31.5	5V630	63.0	8V1250	125.0
3V335	33.5	5V670	67.0	8V1320	132.0
3V355	35.5	5V710	71.0	8V1400	140.0
3V375	37.5	5V750	75.0	8V1500	150.0
3V400	40.0	5V800	80.0	8V1600	160.0
3V425	42.5	5V850	85.0	8V1700	170.0
3V450	45.0	5V900	90.0	8V1800	180.0
3V475	47.5	5V950	95.0	8V1900	190.0
3V500	50.0	5V1000	100.0	8V2000	200.0
3V530	53.0	5V1060	106.0	8V2120	212.0
3V560	56.0	5V1120	112.0	8V2240	224.0
3V600	60.0	5V1180	118.0	8V2360	236.0
3V630	63.0	5V1250	125.0	8V2500	250.0
3V670	67.0	5V1320	132.0	8V2650	265.0
3V710	71.0	5V1400	140.0	8V2800	280.0
3V750	75.0	5V1500	150.0	8V3000	300.0
3V800	80.0	5V1600	160.0	8V3150	315.0
3V850	85.0	5V1700	170.0	8V3350	335.0
3V900	90.0	5V1800	180.0	8V3550	355.0
3V950	95.0	5V1900	190.0	8V3750	375.0
3V1000	100.0	5V2000	200.0	8V4000	400.0
3V1060	106.0	5V2120	212.0	8V4250	425.0
3V1120	112.0	5V2240	224.0	8V4500	450.0
3V1180	118.0	5V2360	236.0	8V5000	500.0
3V1250	128.0	5V2500	250.0		
3V1320	132.0	5V2650	265.0		
3V1400	140.0	5V2800	280.0		
		5V3000	300.0		
		5V3150	315.0		
		5V3350	335.0		
		5V3550	355.0		

If the 60-inch "B" section belt shown is made 3/10 of an inch longer, it will be code marked 53 rather than 50. If made 3/10 shorter, it will be marked 47. While both have the belt number B60 they cannot be used in a set because of the difference in length.

TYPICAL CODE MARKING

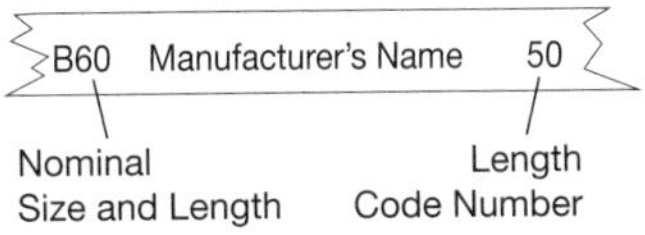

EXTENSION CORD SIZES FOR PORTABLE TOOLS

Cord Length (ft.)	Full-Load Rating of Tool in Amperes at 115 Volts					
	0 to 2.0	2.1 to 3.4	3.5 to 5.0	5.1 to 7.0	7.1 to 12.0	12.1 to 16.0
	Wire Size (AWG)					
25	18	18	18	16	14	14
50	18	18	18	16	14	12
75	18	18	16	14	12	10
100	18	16	14	12	10	8
200	16	14	12	10	8	6
300	14	12	10	8	6	4
400	12	10	8	6	4	4
500	12	10	8	6	4	2
600	10	8	6	4	2	2
800	10	8	6	4	2	1
1,000	8	6	4	2	1	0

CHAPTER 11
Conversion Factors, Mathematics and Units of Measurement

COMMONLY USED CONVERSION FACTORS

Multiply	By	To Obtain
Acres	43,560	Square feet
Acres	1.562×10^{-3}	Square miles
Acre-Feet	43,560	Cubic feet
Amperes per sq cm	6.452	Amperes per sq in.
Amperes per sq in.	0.1550	Amperes per sq cm
Ampere-Turns	1.257	Gilberts
Ampere-Turns per cm . . .	2.540	Ampere-turns per in.
Ampere-Turns per in . . .	0.3937	Ampere-turns per cm
Atmospheres	76.0	Cm of mercury
Atmospheres	29.92	Inches of mercury
Atmospheres	33.90	Feet of water
Atmospheres	14.70	Pounds per sq in.
British thermal units	252.0	Calories
British thermal units	778.2	Foot-pounds
British thermal units	3.960×10^{-4}	Horsepower-hours
British thermal units	0.2520	Kilogram-calories
British thermal units	107.6	Kilogram-meters
British thermal units	2.931×10^{-4}	Kilowatt-hours
British thermal units	1,055	Watt-seconds
B.t.u. per hour	2.931×10^{-4}	Kilowatts
B.t.u. per minute	2.359×10^{-2}	Horsepower
B.t.u. per minute	1.759×10^{-2}	Kilowatts
Bushels	1.244	Cubic feet
Centimeters	0.3937	Inches
Circular mils	5.067×10^{-6}	Square centimeters
Circular mils	0.7854×10^{-6}	Square inches

COMMONLY USED CONVERSION FACTORS *(cont.)*

Multiply	By	To Obtain
Circular mils	0.7854	Square mils
Cords	128	Cubic feet
Cubic centimeters	6.102×10^{-6}	Cubic inches
Cubic feet	0.02832	Cubic meters
Cubic feet	7.481	Gallons
Cubic feet	28.32	Liters
Cubic inches	16.39	Cubic centimeters
Cubic meters	35.31	Cubic feet
Cubic meters	1.308	Cubic yards
Cubic yards	0.7646	Cubic meters
Degrees (angle)	0.01745	Radians
Dynes	2.248×10^{-6}	Pounds
Ergs	1	Dyne-centimeters
Ergs	7.37×10^{-6}	Foot-pounds
Ergs	10^{-7}	Joules
Farads	10^{6}	Microfarads
Fathoms	6	Feet
Feet	30.48	Centimeters
Feet of water	.08826	Inches of mercury
Feet of water	304.8	Kg per square meter
Feet of water	62.43	Pounds per square ft.
Feet of water	0.4335	Pounds per square in.
Foot-pounds	1.285×10^{-2}	British thermal units
Foot-pounds	5.050×10^{-7}	Horsepower-hours
Foot-pounds	1.356	Joules
Foot-pounds	0.1383	Kilogram-meters
Foot-pounds	3.766×10^{-7}	Kilowatt-hours
Gallons	0.1337	Cubic feet
Gallons	231	Cubic inches
Gallons	3.785×10^{-3}	Cubic meters
Gallons	3.785	Liters
Gallons per minute	2.228×10^{-3}	Cubic feet per sec.
Gausses	6.452	Lines per square in.
Gilberts	0.7958	Ampere-turns
Henries	10^{3}	Millihenries
Horsepower	42.41	Btu per min.
Horsepower	2,544	Btu per hour

COMMONLY USED CONVERSION FACTORS *(cont.)*

Multiply	By	To Obtain
Horsepower	550	Foot-pounds per sec.
Horsepower	33,000	Foot-pounds per min.
Horsepower	1.014	Horsepower (metric)
Horsepower	10.70	Kg calories per min.
Horsepower	0.7457	Kilowatts
Horsepower (boiler) ...	33,520	Btu per hour
Horsepower-hours	2,544	British thermal units
Horsepower-hours	1.98×10^{6}	Foot-pounds
Horsepower-hours	2.737×10^{5}	Kilogram-meters
Horsepower-hours	0.7457	Kilowatt-hours
Inches	2.540	Centimeters
Inches of mercury	1.133	Feet of water
Inches of mercury	70.73	Pounds per square ft.
Inches of mercury	0.4912	Pounds per square in.
Inches of water	25.40	Kg per square meter
Inches of water	0.5781	Ounces per square in.
Inches of water	5.204	Pounds per square ft.
Joules	9.478×10^{-4}	British thermal units
Joules	0.2388	Calories
Joules	10^{7}	Ergs
Joules	0.7376	Foot-pounds
Joules	2.778×10^{-7}	Kilowatt-hours
Joules	0.1020	Kilogram-meters
Joules	1	Watt-seconds
Kilograms	2.205	Pounds
Kilogram-calories	3.968	British thermal units
Kilogram meters	7.233	Foot-pounds
Kg per square meter ...	3.281×10^{-3}	Feet of water
Kg per square meter ...	0.2048	Pounds per square ft.
Kg per square meter ...	1.422×10^{-3}	Pounds per square in.
Kilolines	10^{3}	Maxwells
Kilometers	3.281	Feet
Kilometers	0.6214	Miles
Kilowatts	56.87	Btu per min.
Kilowatts	737.6	Foot-pounds per sec.
Kilowatts	1.341	Horsepower
Kilowatts-hours	3409.5	British thermal units

COMMONLY USED CONVERSION FACTORS *(cont.)*

Multiply	By	To Obtain
Kilowatts-hours	2.655×10^6	Foot-pounds
Knots	1.152	Miles
Liters	0.03531	Cubic feet
Liters	61.02	Cubic inches
Liters	0.2642	Gallons
Log N_e or in N	0.4343	Log_{10} N
Log N	2.303	Log_e N or in N
Lumens per square ft.	1	Footcandles
Maxwells	10^{-3}	Kilolines
Megalines	10^6	Maxwells
Megaohms	10^6	Ohms
Meters	3.281	Feet
Meters	39.37	Inches
Meter-kilograms	7.233	Pound-feet
Microfarads	10^{-6}	Farads
Microhms	10^{-6}	Ohms
Microhms per cm cube	0.3937	Microhms per in. cube
Microhms per cm cube	6.015	Ohms per mil foot
Miles	5,280	Feet
Miles	1.609	Kilometers
Miner's inches	1.5	Cubic feet per min.
Ohms	10^{-6}	Megohms
Ohms	10^6	Microhms
Ohms per mil foot	0.1662	Microhms per cm cube
Ohms per mil foot	0.06524	Microhms per in. cube
Poundals	0.03108	Pounds
Pounds	32.17	Poundals
Pound-feet	0.1383	Meter-Kilograms
Pounds of water	0.01602	Cubic feet
Pounds of water	0.1198	Gallons
Pounds per cubic foot	16.02	Kg per cubic meter
Pounds per cubic foot	5.787×10^{-4}	Pounds per cubic in.
Pounds per cubic inch	27.68	Grams per cubic cm
Pounds per cubic inch	2.768×10^{-4}	Kg per cubic meter
Pounds per cubic inch	1.728	Pounds per cubic ft.
Pounds per square foot	0.01602	Feet of water
Pounds per square foot	4.882	Kg per square meter

COMMONLY USED CONVERSION FACTORS *(cont.)*

Multiply	By	To Obtain
Pounds per square foot . . .	6.944×10^{-3}	Pounds per sq. in.
Pounds per square inch . . .	2.307	Feet of water
Pounds per square inch . . .	2.036	Inches of mercury
Pounds per square inch . . .	703.1	Kg per square meter
Radians	57.30	Degrees
Square centimeters	1.973×10^{5}	Circular mils
Square feet	2.296×10^{-5}	Acres
Square feet	0.09290	Square meters
Square inches	1.273×10^{6}	Circular mils
Square inches	6.452	Square centimeters
Square kilometers	0.3861	Square miles
Square meters	10.76	Square feet
Square miles	640	Acres
Square miles	2.590	Square kilometers
Square millimeters	1.973×10^{3}	Circular mils
Square mils	1.273	Circular mils
Tons (long)	2,240	Pounds
Tons (metric)	2,205	Pounds
Tons (short)	2,000	Pounds
Watts	0.05686	Btu per minute
Watts	10^{7}	Ergs per sec.
Watts	44.26	Foot-pounds per min.
Watts	1.341×10^{-3}	Horsepower
Watts	14.34	Calories per min.
Watts-hours	3.412	British thermal units
Watts-hours	2,655	Footpounds
Watts-hours	1.341×10^{-3}	Horsepower-hours
Watts-hours	0.8605	Kilogram-calories
Watts-hours	376.1	Kilogram-meters
Webers	10^{8}	Maxwells

METRIC TO TRADE SIZE

METRIC DESIGNATOR (mm)												
12	16	21	27	35	41	53	63	78	91	103	129	155
3/8	1/2	3/4	1	1 1/4	1 1/2	2	2 1/2	3	3 1/2	4	5	6
TRADE SIZE (inches)												

DECIMAL EQUIVALENTS OF FRACTIONS

8ths	32nds	64ths	64ths
1/8 = .125	1/32 = .03125	1/64 = 0.15625	33/64 = .515625
1/4 = .250	3/32 = .09375	3/64 = .046875	35/64 = .546875
3/8 = .375	5/32 = .15625	5/64 = .078125	37/64 = .57812
1/2 = .500	7/32 = .21875	7/64 = .109375	39/64 = .609375
5/8 = .625	9/32 = .28125	9/64 = .140625	41/64 = .640625
3/4 = .750	11/32 = .34375	11/64 = .171875	43/64 = .671875
7/8 = .875	13/32 = .40625	13/64 = .203128	45/64 = .703125
16ths	15/32 = .46875	15/64 = .234375	47/64 = .734375
1/16 = .0625	17/32 = .53125	17/64 = .265625	49/64 = .765625
3/16 = .1875	19/32 = .59375	19/64 = .296875	51/64 = 3796875
5/16 = .3125	21/32 = .65625	21/64 = .328125	53/64 = .828125
7/16 = .4375	23/32 = .71875	23/64 = .359375	55/64 = .859375
9/16 = .5625	25/32 = .78125	25/64 = .390625	57/64 = .890625
11/16 = .6875	27/32 = .84375	27/64 = .421875	59/64 = .921875
13/16 = .8125	29/32 = .90625	29/64 = .453125	61/64 = .953125
15/16 = .9375	31/32 = .96875	31/64 = .484375	63/64 = .984375

MILLIMETER AND DECIMAL INCH EQUIVALENTS

mm in.	mm in.	mm in.	mm in.	mm in.
1/50 = .00079	30/50 = .02362	11 = .43307	41 = 1.61417	71 = 2.79527
2/50 = .00157	31/50 = .02441	12 = .47244	42 = 1.65354	
3/50 = .00236	32/50 = .02520	13 = .51181	43 = 1.69291	72 = 2.83464
4/50 = .00315	33/50 = .02598	14 = .55118	44 = 1.73228	73 = 2.87401
	34/50 = .02677			74 = 2.91338
5/50 = .00394		15 = .59055	45 = 1.77165	75 = 2.95275
6/50 = .00472	35/50 = .02756	16 = .62992	46 = 1.81102	76 = 2.99212
7/50 = .00551	36/50 = .02835	17 = .66929	47 = 1.85039	
8/50 = .00630	37/50 = .02913	18 = .70866	48 = 1.88976	77 = 3.03149
9/50 = .00709	38/50 = .02992	19 = .74803	49 = 1.92913	78 = 3.07086
	39/50 = .03071			79 = 3.11023
		20 = .78740	50 = 1.96850	80 = 3.14960
10/50 = .00787	40/50 = .03150	21 = .82677	51 = 2.00787	81 = 3.18897
11/50 = .00866	41/50 = .03228	22 = .86614	52 = 2.04724	
12/50 = .00945	42/50 = .03307	23 = .90551	53 = 2.08661	82 = 3.22834
13/50 = .01024	43/50 = .03386	24 = .94488	54 = 2.12598	83 = 3.26771
14/50 = .01102	44/50 = .03465			84 = 3.30708
		25 = .98425	55 = 2.16535	85 = 3.34645
15/50 = .01181	45/50 = .03543	26 = 1.02362	56 = 2.20472	86 = 3.38582
16/50 = .01260	46/50 = .03622	27 = 1.06299	57 = 2.24409	
17/50 = .01339	47/50 = .03701	28 = 1.10236	58 = 2.28346	87 = 3.42519
18/50 = .01417	48/50 = .03780	29 = 1.14173	59 = 2.32283	88 = 3.46456
19/50 = .01496	49/50 = .03858			89 = 3.50393
		30 = 1.18110	60 = 2.36220	90 = 3.54330
20/50 = .01575	1 = .03937	31 = 1.22047	61 = 2.40157	91 = 3.58267
21/50 = .01654	2 = .07874	32 = 1.25984	62 = 2.44094	
22/50 = .01732	3 = .11811	33 = 1.29921	63 = 2.48031	92 = 3.62204
23/50 = .01811	4 = .15748	34 = 1.33858		93 = 3.66141
24/50 = .01890			64 = 2.51968	94 = 3.70078
	5 = .19685	35 = 1.37795	65 = 2.55905	95 = 3.74015
	6 = .23622	36 = 1.41732	66 = 2.59842	96 = 3.77952
25/50 = .01969	7 = .27559	37 = 1.45669		
26/50 = .02047	8 = .31496	38 = 1.49606	67 = 2.63779	97 = 3.81889
27/50 = .02126	9 = .35433	39 = 1.53543	68 = 2.67716	98 = 3.85826
28/50 = .02205			69 = 2.71653	99 = 3.89763
29/50 = .02283	10 = .39370	40 = 1.57480	70 = 2.75590	100 = 3.93700

METRIC MEASUREMENTS

Prefixes

Mega	=	1,000,000	Deci	=	0.1
Kilo	=	1,000	Centi	=	0.01
Hecto	=	100	Milli	=	0.001
Deka	=	10	Micro	=	0.000001

Linear Measures

(1 Meter = 39.37 Inches)

1 Centimeter	=	10 Millimeters	=	0.3937	In.
1 Decimeter	=	10 Centimeters	=	3.937	Ins.
1 Meter	=	10 Decimeters	=	1.0936	Yds.
1 Dekameter	=	10 Meters	=	10.9361	Yds.
1 Hectometer	=	10 Dekameters	=	109.3614	Yds.
1 Kilometer	=	10 Hectometers	=	0.6213	Mile

Square Measures

(1 Square Meter = 1549.997 Sq. Inches)

1 Sq. Centimeter	=	100 Sq. Millimeters	=	0.155	Sq. In.
1 Sq. Decimeter	=	100 Sq. Centimeters	=	15.55	Sq. Ins.
1 Sq. Meter	=	100 Sq. Decimeters	=	10.764	Sq. Ft.
1 Sq. Dekameter	=	100 Sq. Meters	=	119.6	Sq. Yds.
1 Sq. Hectometer	=	100 Sq. Dekameters			
1 Sq. Kilometer	=	100 Sq. Hectometers			

(1 Are = 100 Sq. Meters)

1 Centiare	=	10 Milliares	=	10.764	Sq. Ft.
1 Deciare	=	10 Centiares	=	11.96	Sq. Yds.
1 Are	=	10 Deciares	=	119.6	Sq. Yds.
1 Dekare	=	10 Ares	=	0.247	Acres
1 Hektare	=	10 Dekares	=	2.471	Acres
1 Sq. Kilometer	=	100 Hektares	=	0.386	Sq. Mile

METRIC MEASUREMENTS *(cont.)*

Cubic Measures

(1 Meter = 39.37 Inches)

1 Cu. Centimeter	=	1000 Cu. Millimeters	=	0.061 Cu. In.
1 Cu. Decimeter	=	1000 Cu. Centimeters	=	61.023 Cu. In.
1 Cu. Meter	=	1000 Cu. Decimeters	=	35.314 Cu. Ft.
1 Cu. Meter	=	1 Stere	=	1.3079 Cu. Yds.
1 Cu. Centimeter			=	1 Gram
1000 Cu. Centimeters (1 Liter)			=	1 Kilogram
1 Cu. Meter (1000 Liters)			=	1 Metric Ton

Weight Measures

(1 Gram = 0.0352 Ounces)

1 Centigram	=	10 Milligrams	=	0.154	Grains
1 Decigram	=	10 Centigrams	=	1.543	Grains
1 Gram	=	10 Decigrams	=	15.432	Grains
1 Dekagram	=	10 Grams	=	5.643	Drams
1 Hectogram	=	10 Dekagrams	=	3.527	Ounces
1 Kilogram	=	10 Hectograms	=	2.2046	Pounds
1 Myriagram	=	10 Kilograms	=	22.0462	Pounds
1 Quintal	=	10 Myriagrams	=	220.4622	Pounds
1 Metric Ton	=	10 Quintal	=	2,204.622	Pounds

Capacity Measures

(1 Liter = 1.0567 Liquid Quarts)

1 Centiliter	=	10 Milliliters	=	0.338	Fluid Ounces
1 Deciliter	=	10 Centiliters	=	3.38	Fluid Ounces
1 Liter	=	10 Deciliters	=	33.8	Fluid Ounces
1 Dekaliter	=	10 Liters	=	0.284	Bushel
1 Hectoliter	=	10 Dekaliters	=	2.84	Bushels
1 Kiloliter	=	10 Hectoliters	=	264.2	Gallons

U.S. MEASUREMENTS

Linear Measures

			1	Inch	=	2.540	Centimeters
12	Inches	=	1	Foot	=	3.048	Decimeters
3	Feet	=	1	Yard	=	9.144	Decimeters
5.5	Yards	=	1	Rod	=	5.029	Meters
40	Rods	=	1	Furlong	=	2.018	Hectometers
8	Furlongs	=	1	Mile	=	1.609	Kilometers
1	Stature Mile	=	5,280	Feet			
1	Span	=	9	Inches			
1	Chain	=	22	Yards			
1	Link	=	7.92	Inches			
1	Fathom	=	6	Feet			
1	Furlong	=	10	Chains			
1	Cable	=	608	Feet			

Square Measures

144	Square Inches	=	1	Square Foot
9	Square Feet	=	1	Square Yard
30.25	Square Yards	=	1	Square Rod
40	Rods	=	1	Rood
4	Roods	=	1	Acre
640	Acres	=	1	Square Mile
1	Square Mile	=	1	Section
36	Sections	=	1	Township

Cubic or Solid Measures

1	Cu. Foot	=	1728	Cu. Inches
1	Cu. Yard	=	27	Cu. Feet
1	Cu. Foot	=	7.48	Gallons
1	Gallon (Water)	=	8.34	Lbs.
1	Gallon (U.S.)	=	231	Cu. Inches of Water
1	Gallon (Imperial)	=	277.25	Cu. Inches of Water

U.S. MEASUREMENTS *(cont.)*

Liquid Measures

1	U.S. Pint	=	16	Fl. Ounces
1	Standard Cup	=	8	Fl. Ounces
1	Tablespoon	=	0.5	Fl. Ounces
1	Teaspoon	=	0.16	Fl. Ounces
1	Pint	=	2	Cups
1	Quart	=	2	Pints
1	Gallon	=	4	Quarts
1	Barrel	=	42	Gallons (Crude Oil)

Dry Measures

1	Bushel	=	8	Gallons

Weight (Mass) Measures

Avoirdupois Weight:

1	Ounce	=	16	Drams
1	Pound	=	16	Ounces
1	Hundredweight	=	100	Pounds
1	Ton	=	2,000	Pounds

Troy Weight:

1	Carat	=	3.17	Grains
1	Pennyweight	=	20	Grains
1	Ounce	=	20	Pennyweights
1	Pound	=	12	Ounces
1	Long Hundred-weight	=	112	Pounds
1	Long Ton	=	2240	Pounds

Apothecaries Weight:

1	Scruple	=	20	Grains	=	1.296	Grams
1	Dram	=	3	Scruples	=	3.888	Grams
1	Ounce	=	8	Drams	=	31.10	Grams
1	Pound	=	12	Ounces	=	373.24	Grams

MATHEMATICS

Formulas

Formulas are statements of truth. If we say that $X = 1 + 2$, we can then manipulate the formula to say $X - 2 = 1$, or, $X - 1 = 2$. This is called *transposing* a formula. (Obviously, $X = 3$.)

Add or Subtract from Both Sides

Formulas are balanced statements of equality. Changing one side and not the other puts it out of balance, but changing both sides equally is fine.

Subtract 5 from each side: $x = y + 5$

$x - 5 = y + 5 - 5$ Since $+5 - 5 = 0$, $x - 5 = y$

$a = b - 5$

Add 5 to each side:

$a + 5 = b - 5 + 5$ $a + 5 = b$

Multiplying or Dividing Both Sides

You can multiply or divide, so long as you do it equally to both sides.

$x = y \div 4$ Multiply both sides by 4:

$4x = y \div 4 \times 4$, Since $\div 4$ and $\times 4$ cancel, $4x = y$

$a = 4b$ Divide each side by 4: $a \div 4 = 4b \div 4$

since $\times b$ and $\div b$ cancel, $a \div 4 = b$

Performing Operations

You can perform other operations to each side of a formula without throwing it out of balance.

For example, we can square each side of an equation without ruining its effectiveness as a true statement. Here is an example of this:

$c = \sqrt{d}$ Squaring both sides: $c^2 = d$

Square Roots

A *square* is any number multiplied by itself.

We note a square with a superscript. "5 squared" = 5^2.

In addition to squares (also called "to the second power"), we can raise numbers to higher powers as well. "5 to the third power" is written 5^3, and would equal $5 \times 5 \times 5$, or 125.

A square root is the number that, multiplied by itself, would equal the number in question. The $\sqrt{\ }$ symbol denotes square root.

Fractions

Fractions are statements of division. The top portion of the fraction is called the numerator and the bottom portion is called the denominator.

$$\frac{\text{numerator}}{\text{denominator}}$$

For this: $\frac{2}{3}$,

we can say "2 over 3" or "2 divided by 3."

$\frac{2}{3} = 2 \div 3 = .67$

Adding and Subtracting Fractions

Adding fractions that have the same denominator is simple. The denominators stay the same, and we simply add or subtract the numerators.

$\frac{1}{8} + \frac{4}{8} + \frac{5}{8} = \frac{10}{8}$ or $1\frac{2}{8}$

$\frac{7}{16} - \frac{2}{16} = \frac{5}{16}$

If the denominators are not equal, we first create identical denominators.

$\frac{1}{3} + \frac{1}{4}$

$\frac{1}{3} = \frac{4}{12}$ $\qquad$ $\frac{1}{4} = \frac{3}{12}$

$\frac{4}{12} + \frac{3}{12} = \frac{7}{12}$, thus $\frac{1}{3} + \frac{1}{4} = \frac{7}{12}$

MATHEMATICS *(cont.)*

Multiplying Fractions

To multiply fractions you multiply the numerator by the numerator and the denominator by the denominator:

$$\frac{2}{3} \times \frac{1}{2} = \frac{2}{6} = \frac{1}{3}$$

Dividing Fractions

Dividing fractions is exactly the same as multiplication, except you invert the second fraction before multiplying, as shown here:

$$\frac{2}{3} \div \frac{1}{2} = \frac{2}{3} \times \frac{2}{1} = \frac{4}{3}$$

Decimals

Decimals are actually a type of fraction that always have a multiple of ten in their denominator.

$$\frac{42}{10} = 4.2$$

$$\frac{5}{100} = .05$$

$$\frac{28}{100} = .28$$

When multiplying numbers by factors of 10, the decimal point is moved to the right.

$51.45 \times 10 = 514.5$
$51.45 \times 100 = 5{,}145.0$
$51.45 \times 1000 = 51{,}450.0$

When dividing by factors of 10, the decimal point is moved to the left.

$62.56 \div 10 = 6.256$
$62.56 \div 100 = .6256$
$62.56 \div 1000 = .06256$

Scientific Notation

Scientific notation (also referred to as *exponents*) is a compact method of expressing very large or very small numbers. Numbers can be expressed as 1 through 10 multiplied by a power of ten. Values above 10 have positive exponents, values less than 1 have negative exponents.

$420{,}000 = 4.2 \times 10^5$
$.00214 = 2.14 \times 10^{-3}$

The Pythagorean Theorem

Describes the relationship of the sides of a **right triangle.** A right triangle always has one 90 degree angle. The longest side of a triangle is always called the **hypotenuse.**

The Pythagorean Theorem states:

$$C^2 = A^2 + B^2, \qquad \text{or,} \qquad C = \sqrt{A^2 + B^2}$$

This formula is true for every right triangle, and it is very useful. A simple right triangle is one with side measurements of 3, 4, and 5.

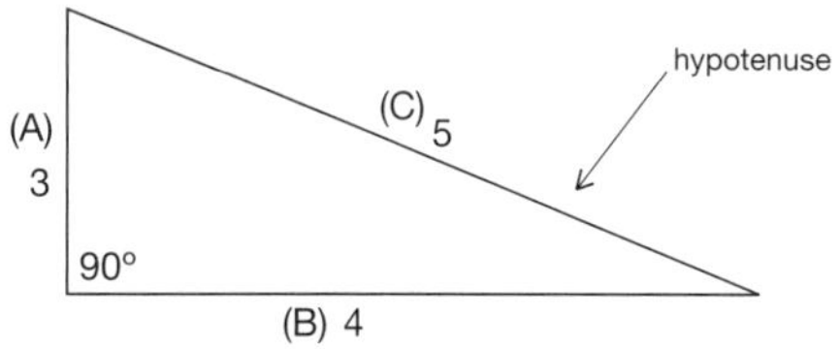

Trigonometric Functions

The primary functions are sine, cosine, and tangent, which state the relationship between the sides of a right triangle.

$$\text{sine} = \frac{\text{opposite}}{\text{hypotenuse}}$$

$$\text{cosine} = \frac{\text{adjacent}}{\text{hypotenuse}}$$

$$\text{tangent} = \frac{\text{opposite}}{\text{adjacent}}$$

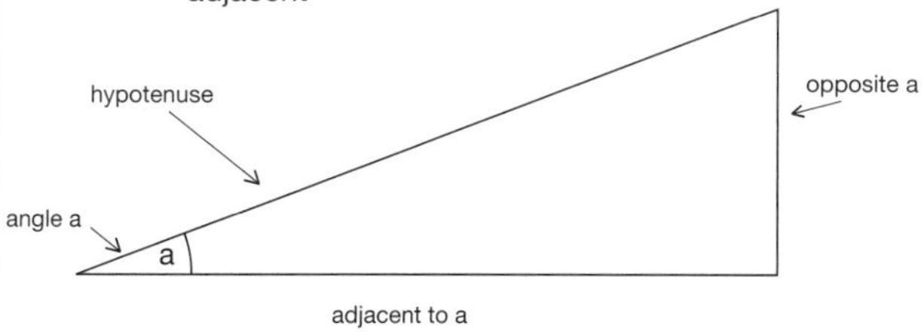

COMMONLY USED GEOMETRICAL RELATIONSHIPS

Diameter of a circle × 3.1416 = Circumference

Radius of a circle × 6.283185 = Circumference

Square of the radius of a circle × 3.1416 = Area

Square of the diameter of a circle × 0.7854 = Area

Square of the circumference of a circle × 0.07958 = Area

Half the circumference of a circle × half its diameter = Area

Circumference of a circle × 0.159155 = Radius

Square root of the area of a circle × 0.56419 = Radius

Circumference of a circle × 0.31831 = Diameter

Square root of the area of a circle × 1.12838 = Diameter

Diameter of a circle × 0.866 = Side of an inscribed equilateral triangle

Diameter of a circle × 0.7071 = Side of an inscribed square

Circumference of a circle × 0.225 = Side of an inscribed square

Circumference of a circle × 0.282 = Side of an equal square

Diameter of a circle × 0.8862 = Side of an equal square

Base of a triangle × one-half the altitude = Area

Multiplying both diameters and .7854 together = Area of an ellipse

Surface of a sphere × one-sixth of its diameter = Volume

Circumference of a sphere × its diameter = Surface

Square of the diameter of a sphere × 3.1416 = Surface

Square of the circumference of a sphere × 0.3183 = Surface

Cube of the diameter of a sphere × 0.5236 = Volume

Cube of the circumference of a sphere × 0.016887 = Volume

Radius of a sphere × 1.1547 = Side of an inscribed cube

Diameter of a sphere divided by $\sqrt{3}$ = Side of an inscribed cube

Area of its base × one-third of its altitude = Volume of a cone or pyramid whether round, square or triangular

Area of one of its sides × 6 = Surface of the cube

Altitude of trapezoid × one-half the sum of its parallel sides = Area

COMMON ENGINEERING UNITS AND THEIR RELATIONSHIP			
Quantity	**SI Metric Units/Symbols**	**Customary Units**	**Relationship of Units**
Acceleration	meters per second squared (m/s^2)	feet per second squared (ft/s^2)	$m/s^2 = ft/s^2 \times 3.281$
Area	square meter (m^2) square millimeter (mm^2)	square foot (ft^2) square inch (in^2)	$m^2 = ft^2 \times 10.764$ $mm^2 = in^2 \times 0.00155$
Density	kilograms per cubic meter (kg/m^3) grams per cubic centimeter (g/cm^3)	pounds per cubic foot (lb/ft^3) pounds per cubic inch (lb/in^3)	$kg/m^3 = lb/ft^2 \times 16.02$ $g/cm^3 = lb/in^3 \times 0.036$
Work	Joule (J)	foot pound force (ft lbf or ft lb)	$J = ft\ lbf \times 1.356$
Heat	Joule (J)	British thermal unit (Btu) Calorie (Cal)	$J = Btu \times 1.055$ $J = cal \times 4.187$
Energy	kilowatt (kW)	Horsepower (HP)	$kW = HP \times 0.7457$
Force	Newton (N) Newton (N)	Pound-force (lbf, lb · f, or lb) kilogram-force (kgf, kg · f, or kp)	$N = lbf \times 4.448$ $N = \frac{kgf}{9.807}$
Length	meter (m) millimeter (mm)	foot (ft) inch (in)	$m = ft \times 3.281$ $mm = \frac{in}{25.4}$
Mass	kilogram (kg) gram (g)	pound (lb) ounce (oz)	$kg = lb \times 2.2$ $g = \frac{oz}{28.35}$
Stress	Pascal = Newton per second (Pa = N/s)	pounds per square inch (lb/in^2 or psi)	$Pa = lb/in^2 \times 6{,}895$
Temperature	degree Celsius (°C)	degree Fahrenheit (°F)	$°C = \frac{°F - 32}{1.8}$
Torque	Newton meter (N · m)	foot-pound (ft lb) inch-pound (in lb)	$N \cdot m = ft\ lbf \times 1.356$ $N \cdot m = in\ lbf \times 0.113$
Volume	cubic meter (m^3) cubic centimeter (cm^3)	cubic foot (ft^3) cubic inch (in^3)	$m^3 = ft^3 \times 35.314$ $cm^3 = \frac{in^3}{16.387}$

CONVERSION TABLE FOR TEMPERATURE – °F/°C

°F	°C	°F	°C	°F	°C	°F	°C	°F	°C
−459.4	−273	−22.0	−30	35.6	2	93.2	34	150.8	66
−418.0	−250	−18.4	−28	39.2	4	96	36	154.4	68
−328.0	−200	−14.8	−26	42.8	6	100.4	38	158.0	70
−238.0	−150	−11.2	−24	46.4	8	104.0	40	161.6	72
−193.0	−125	−7.6	−22	50.0	10	107.6	42	165.2	74
−148.0	−100	−4.0	−20	53.6	12	111.2	44	168.8	76
−130.0	−90	−0.4	−18	57.2	14	114.8	46	172.4	78
−112.0	−80	3.2	−16	60.8	16	118.4	48	176.0	80
−94.0	−70	6.8	−14	64.4	18	122.0	50	179.6	82
−76.0	−60	10.4	−12	68.0	20	125.6	52	183.2	84
−58.0	−50	14.0	−10	71.6	22	129.2	54	186.8	86
−40.0	−40	17.6	−8	75.2	24	132.8	56	190.4	88
−36.4	−38	21.2	−6	78.8	26	136.4	58	194.0	90
−32.8	−36	24.8	−4	82.4	28	140.0	60	197.6	92
−29.2	−34	28.4	−2	86.0	30	143.6	62	201.2	94
−25.6	−32	32.0	0	89.6	32	147.2	64	204.8	96

1 degree F is 1/180 of the difference between the temperature of melting ice and boiling water.
1 degree C is 1/100 of the difference between the temperature of melting ice and boiling water.

Absolute Zero = 273.16°C = −459.69°F

CONVERSION TABLE FOR TEMPERATURE – °F/°C *(cont.)*

°F	°C	°F	°C	°F	°C	°F	°C	°F	°C
208.4	98	347.0	175	590	310	1004	540	6332	3500
212.0	100	356.0	180	608	320	1040	560	7232	4000
221.0	105	365.0	185	626	330	1076	580	4500	8132
230.0	110	374.0	190	644	340	1112	600	9032	5000
239.0	115	383.0	195	662	350	1202	650	9932	5500
248.0	120	392.0	200	680	360	1292	700	10832	6000
257.0	125	410	210	698	370	1382	750	11732	6500
266.0	130	428	220	716	380	1472	800	12632	7000
275.0	135	446	230	734	390	1562	850	13532	7500
284.0	140	464	240	752	400	1652	900	14432	8000
293.0	145	482	250	788	420	1742	950	15332	8500
302.0	150	500	260	824	440	1832	1000	16232	9000
311.0	155	518	270	860	460	2732	1500	17132	9500
320.0	160	536	280	896	480	3632	2000	18032	10000
329.0	165	554	290	932	500	4532	2500		
338.0	170	572	300	968	520	5432	3000		

1 degree F is 1/180 of the difference between the temperature of melting ice and boiling water.
1 degree C is 1/100 of the difference between the temperature of melting ice and boiling water.

Absolute Zero = 273.16°C = −459.69°F

ELECTRICAL PREFIXES

Prefixes

Prefixes are used to avoid long expressions of units that are smaller and larger than the base unit. See Common Prefixes. For example, sentences 1 and 2 do not use prefixes. Sentences 3 and 4 use prefixes.

1. A solid-state device draws 0.000001 amperes (A).
2. A generator produces 100,000 watts (W).
3. A solid-state device draws 1 microampere (μA).
4. A generator produces 100 kilowatts (kW).

Converting Units

To convert between different units, the decimal point is moved to the left or right, depending on the unit. See Conversion Table. For example, an electronic circuit has a current flow of .000001 A. The current value is converted to simplest terms by moving the decimal point six places to the right to obtain 1.0μA (from Conversion Table).

.000001. A = 1.0 μA

Move decimal point
6 places to right

Common Electrical Quantities

Abbreviations are used to simplify the expression of common electrical quantities. See Common Electrical Quantities. For example, milliwatt is abbreviated mW, kilovolt is abbreviated kV, and ampere is abbreviated A.

COMMON PREFIXES

Symbol	Prefix	Equivalent
G	giga	1,000,000,000
M	mega	1,000,000
k	kilo	1000
base unit	—	1
m	milli	.001
u	micro	.000001
n	nano	.000000001

COMMON ELECTRICAL QUANTITIES

Variable	Name	Unit of Measure and Abbreviation
E	voltage	volt - V
I	current	ampere - A
R	resistance	ohm - Ω
P	power	watt - W
P	power (apparent)	volt-amp - VA
C	capacitance	farad - F
L	inductance	henry - H
Z	impedance	ohm - Ω
G	conductance	siemens - S
f	frequency	hertz - Hz
T	period	second - s

CONVERSION TABLE

Initial Units	Final Units: giga	mega	kilo	base unit	milli	micro	nano
giga	—	3R	6R	9R	12R	15R	18R
mega	3L	—	3R	6R	9R	12R	15R
kilo	6L	3L	—	3R	6R	9R	12R
base unit	9L	6L	3L	—	3R	6R	9R
milli	12L	9L	6L	3L	—	3R	6R
micro	15L	12L	9L	6L	3L	—	3R
nano	18L	15L	12L	9L	6L	3L	—

CHAPTER 12
Symbols

SWITCH OUTLETS	
S	Single-pole switch
S_2	Double-pole switch
S_3	Three-way switch
S_4	Four-way switch
S_K	Key-operated switch
S_P	Switch and pilot lamp
S_{wp}	Weather-proof switch
S_F	Fused switch
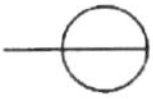 S	Switch and single receptacle
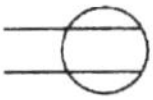 S	Switch and duplex receptacle

SWITCH OUTLETS *(cont.)*

S_D — Door switch

S_T — Time switch

S_{CB} — Circuit-breaker switch

S_{MC} — Momentary contact switch or push-button for other than signaling system

LIGHTING OUTLETS

CEILING **WALL**

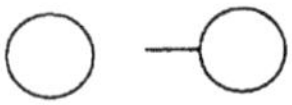

Surface or pendant incandescent, mercury vapor, or similar lamp fixture

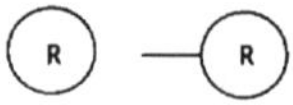

Recessed incandescent, mercury vapor, or similar lamp fixture

Surface or pendant individual fluorescent fixture

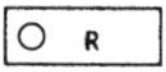

Recessed individual fluorescent fixture

LIGHTING OUTLETS *(cont.)*

Surface or pendant continuous-row fluorescent fixture

Recessed continuous-row fluorescent fixture*

Bare-lamp fluorescent strip**

CEILING **WALL**

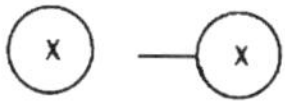

Surface or pendant exit light

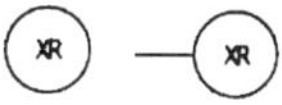

Recessed exit light

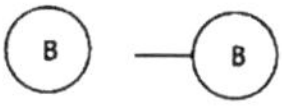

Blanked outlet

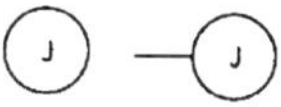

Junction box

* *In the case of combination continuous-row fluorescent and incandescent spotlights, use combinations of the above Standard symbols.*

** *In the case of a continuous-row bare-lamp fluorescent strip above an area-wide diffusion means, show each fixture run, using the Standard symbol; indicate area of diffusing means and type of light shading and/or drawing notation.*

RECEPTACLE OUTLETS

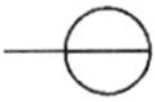	Single receptacle outlet
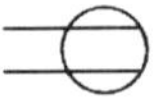	Duplex receptacle outlet
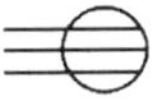	Triplex receptacle outlet
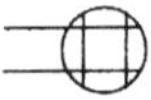	Quadruplex receptacle outlet
	Duplex receptacle outlet-split wired
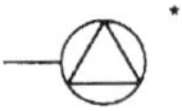	Single special-purpose receptacle outlet*
	Duplex special-purpose receptacle outlet*
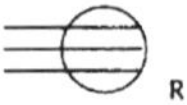	Range outlet

Unless noted, assume every receptacle will be grounded, and will have a separate grounding contact.

** Use numeral or letter, either within the symbol or as a subscript alongside the symbol keyed to explanation in the drawing list of symbols, to indicate type of receptacle or usage.*

Special-purpose connection or provision for connection. Use subscript letters to indicate function (DW—dishwasher; CD—clothes dryer, etc.)

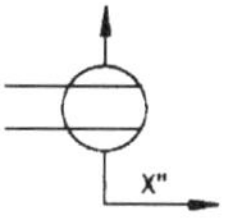

Multioutlet assembly. Extend arrows to limit of installation. Use appropriate symbol to indicate type of outlet. Also indicate spacing of outlets as × inches.

Pushbutton

Buzzer

Bell

Chime

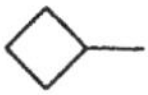

Annunciator

Electric door opener

Telephone outlet

Television outlet

Unless noted, assume every receptacle will be grounded, and will have a separate grounding contact.

* *Use numeral or letter, either within the symbol or as a subscript alongside the symbol keyed to explanation in the drawing list of symbols, to indicate type of receptacle or usage.*

RECEPTACLE OUTLETS *(cont.)*

	Clock hanger receptacle
	Fan hanger receptacle
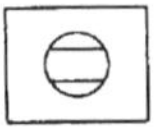	Floor outlet, duplex receptacle
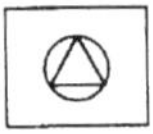	Floor outlet*, special-purpose
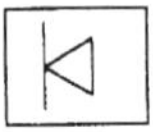	Floor outlet, telephone

GENERAL

	Electric motor
	Electric generator
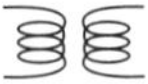	Transformer
	Watthour meter

Unless noted, assume every receptacle will be grounded, and will have a separate grounding contact.

* *Use numeral or letter, either within the symbol or as a subscript alongside the symbol keyed to explanation in the drawing list of symbols, to indicate type of receptacle or usage.*

GENERAL *(cont.)*

Circuit breaker

Fusible element

Single-throw knife switch

Double-throw knife switch

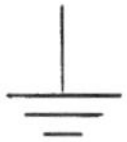

Ground

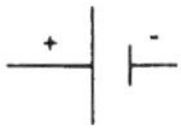

Battery

GENERAL *(cont.)*

Manhole

Handhole

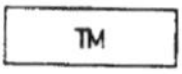

Transformer manhole or vault

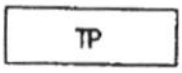

Transformer pad

Underground direct burial cable. Indicate type, size, and number of conductors by notation or schedule.

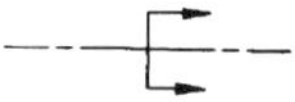

Underground duct line. Indicate type, size, and number of ducts by cross-section identification of each run by notation or schedule. Indicate type, size, and number of conductors by notation or schedule.

Streetlight standard feed from underground circuit

PANELBOARDS, SWITCHBOARDS AND RELATED EQUIPMENT

	Flush-mounted panelboard or cabinet
	Surface-mounted panelboard or cabinet
	Switchboard, power control center, unit substations – should be drawn to scale
	Motor or other power controller
	Externally-operated disconnection switch
	Combination controller and disconnection means

CIRCUITING AND WIRING

	Wiring concealed in ceiling or wall
	Wiring concealed in floor
	Wiring exposed
 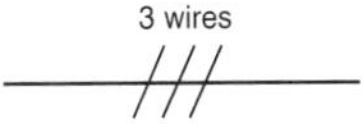	Branch-circuit home run to panelboard. Number of arrows indicates number of circuits. Any circuit without further identification indicates two-wire circuit. For a greater number of wires, indicate with cross lines.

CIRCUITING AND WIRING *(cont.)*

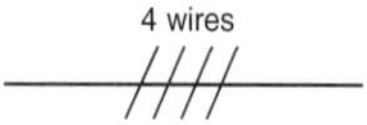	Unless indicated otherwise, the wire size of the circuit is the minimum size required by the specification.
	Wiring turned up
	Wiring turned down

MOTOR CONTROL

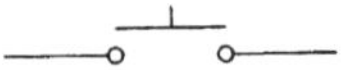	Momentary-contact switch, normally open
	Momentary-contact switch, normally closed
	Inductor, Iron-core
	Armature
	Crossed wires, not connected
	Crossed wires, connected
	Fuse

MOTOR CONTROL *(cont.)*

Symbol	Description
	Limit switch, normally open
	Limit switch, normally closed
	Footswitch, normally open
	Footswitch, normally closed
	Vacuum switch, normally open
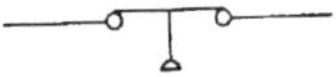	Vacuum switch, normally closed
	Liquid-level switch, normally open
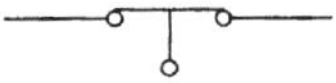	Liquid-level switch, normally closed

MOTOR CONTROL *(cont.)*

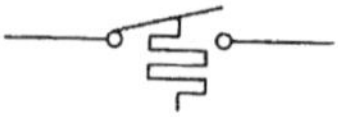

Temperature-actuated switch, normally open

Temperature-actuated switch, normally closed

Flow switch, normally open

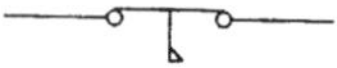

Flow switch, normally closed

Magnetic blowout coil

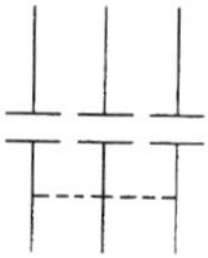

Manually operated 3-pole contactor

Thermocouple

Diode (rectifier)

Capacitor

MOTOR CONTROL *(cont.)*

Adjustable capacitor

Resistor

Tapped resistor

Variable resistor

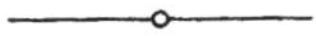

Wiring terminal

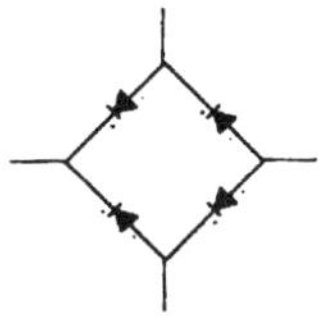

Full-wave rectifier

Mechanical interlock

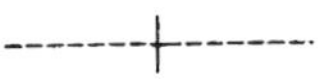

Mechanical connection

SWITCH

Switch symbols are constructed of basic symbols for mechanical connections, contacts, etc. and normally a switch is represented in the de-energized position for switches having two or more positions where no operating force is applied. When actuated by a mechanical force, the functioning point is described by a clarifying note. Where switch symbols are in closed position, the terminals should be included for clarity.

Single throw, general

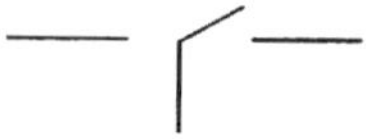

Double throw, general

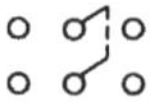

2-pole double throw switch with terminals shown

Knife switch, general

PUSHBUTTON, MOMENTARY OR SPRING RETURN

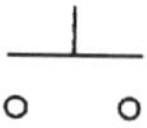

Circuit closing (make)

Circuit opening (break)

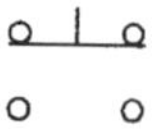

Two-circuit

PUSHBUTTON, MAINTAINED OR NOT SPRING RETURN

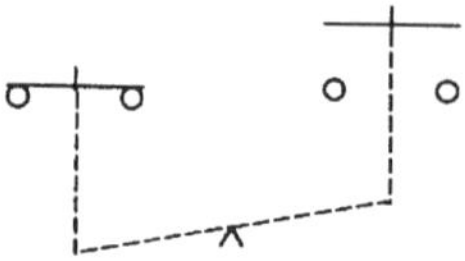

Two-circuit

CHAPTER 13
Glossary

GLOSSARY

10Base2: 10 Mbps, baseband, in 185 meter segments. The IEEE 802.3 substandard for ThinWire, coaxial, Ethernet.

10BaseT: 10 Mbps, baseband, over twisted pair. The IEEE 802.3 substandard for unshielded twisted pair Ethernet.

110 Type Block: A wire connecting block that terminates 100 to 300 pairs of wire.

66-Type Block: A type of wire connecting block that is used for twisted pair cabling cross connections. It holds 25 pairs in up to four columns.

AC Current: Electrical current which reverses direction at regular intervals (cycles) due to a change in voltage which occurs at the same frequency.

AC Line Filter: Absorbs electrical interference from motors, etc.

AC Sine Wave: A symmetrical waveform that contains 360 electrical degrees.

AC Voltage: Electrical pressure that reverses direction at regular intervals called cycles.

Absorption: Loss of power in an optical fiber, resulting from conversion of optical power into heat and caused principally by impurities, such as transition metals and hydroxyl ions.

Access Line: A line or circuit that connects a customer site to a network switching center or local exchange. Also known as the local loop.

Aerial Cable: Telecommunications cable installed on aerial supporting structures such as poles, sides of buildings, and other structures.

Alternator: Converts mechanical energy into electrical energy.

Ambient Temperature: The temperature of the surroundings.

American National Standards Institute (ANSI): A private organization that coordinates some U.S. standards setting.

GLOSSARY *(cont.)*

American Standard Code for Information Interchange (ASCII): A standard character set that (typically) assigns a 7-bit sequence to each letter, number, and selected control character.

American Wire Gauge (AWG): Standard used to describe the size of a wire. The larger the AWG number, the smaller (thinner) the described wire.

Ammeter: An instrument (meter) for measuring electrical current.

Ampacity: The amount of current (measured in amperes) that a conductor can carry without overheating.

Ampere (A): Unit of current measurement. The amount of current that will flow through a one ohm resistor when one volt is applied.

Ampere-hour: The quantity of electricity equal to the flow of a current of one ampere for one hour.

Amplitude: The size, in voltage, of signals in a data transmission.

Angular Misalignment: The loss of optical power caused by deviation from optimum alignment of fiber to fiber or fiber to waveguide.

Annunciator: A sound generating device that intercepts and speaks the condition of circuits or circuits' operations.

Anode: The positive electrode in an electrochemical cell (battery) toward which current flows.

Apparent Power (P_A): Product of the voltage and current in a circuit calculated without considering the phase shift that might be present between the voltage and the current. Expressed in terms of Volt: Ampere (VA).

Approved Ground: A grounding bus or strap in a building that is suitable for connecting to data communication equipment.

Arcing: A luminous discharge formed by the span of electrical current across a space between electrodes or conductors.

Arc Tube: The light-producing element of an HID lamp.

Armature: The rotating part of a DC motor.

Arrestor (lightning): A device that reduces the voltage of a surge applied to its terminals and restores itself to its original operating condition.

Attenuation: Denotes the loss in strength of power between that transmitted and that received. Usually expressed as a ratio in dB (decibel).

Attenuator: A device that reduces signal power in a fiber optic link by inducing loss.

Autotransformer: Changes voltage level using the same common coil for both the primary and the secondary.

Avalanche Current: Current passed when a diode malfunctions.

Average Value (Vavg): The mathematical mean of all instantaneous voltage values in a sine wave.

B: Byte.

b: bit.

Back Electromagnetic Force: The voltage created in an inductive circuit by a changing current flowing through the circuit.

Back Reflection, Optical Return Loss: Light reflected from the polished end of a fiber caused by the difference of refractive indices.

Backboard: A wooden (or metal) panel used for mounting equipment usually on a wall.

Backbone: The main connectivity device of a distributed system. All systems that have connectivity to the backbone will connect to each other.

Ballast: An electrical circuit component used with fluorescent lamps to provide the voltage necessary to strike the mercury arc within the lamp, and then to limit the amount of current that flows through the lamp.

Bandwidth: Technically, the difference, in Hertz (Hz), between the highest and lowest frequencies of a transmission channel.

Bank: An assemblage of fixed contacts.

Battery: A device that converts chemical energy into electrical current.

GLOSSARY *(cont.)*

Baud: A unit of signaling speed. The speed in Baud is the number of discrete conditions or signal elements per second.

Bidirectional Current: Has both positive and negative values.

Bits Per Second (bps): Basic unit of measurement for serial data transmission capacity, abbreviated as k bps, or kilobit/s, for thousands of bits per second; m bps, or megabit/s, for millions of bits per second; g bits, or gigabit/s for billions of bits per second.

Blocking Diode: A diode used to prevent current flow in a photovoltaic array during times when the array is not producing electricity.

Bonding: A very low impedance path accomplished by permanently joining non-current-carrying metal parts. It is done to provide electrical continuity and to offer the capacity to safely conduct any current.

Bonding Jumper: A conductor used to assure the required electrical connection between metal parts of an electrical system.

Bonding Conductor: The conductor that connects the non-current-carrying parts of electrical equipment, cable raceways, or other enclosures to the approved system ground conductor.

Bond Wire: Bare grounding wire that runs inside of an armored cable.

Branch Circuit: Conductors between the last overcurrent device and the outlets.

Branch Circuit, Multiwire: A branch circuit having two or more ungrounded circuit conductors, each having a voltage difference between them, and a grounded circuit conductor (neutral) having an equal voltage difference between it and each ungrounded conductor.

Break: The number of separate places on a contact that open or close a circuit.

Breakout Box: A device that allows access to individual points on a physical interface connector for testing and monitoring.

Breakout Cable: A multifiber cable where each fiber is protected by an additional jacket and strength element beyond the overall cable.

GLOSSARY *(cont.)*

British Thermal Unit (BTU): The amount of heat necessary to raise the temperature of one pound of water 1°F.

Brushes: Sliding contacts that provide the connection between the external power circuit and the commutator of a DC motor.

Bus: A group of conductors that serve as a common connection for circuits.

Bus Bar: The heavy copper or aluminum bar used to carry currents in switchboards.

Busway: A metal enclosed distribution of bus bars.

Cable: One or more insulated or non-insulated wires used to conduct electrical current.

Calorie (Cal): The amount of heat required to raise 1 gallon of water 1°C.

Capacitance (C): The ability of a circuit or component to store an electrical charge and is measured in farads (F).

Capacitive Circuit: A circuit in which current leads voltage (voltage lags current).

Capacitive Reactance (Xc): The opposition to current flow by a capacitor measured in Ohms (Ω).

Capacitor: A device that stores electrical energy by means of an electrostatic field.

Cathode: The negative electrode in an electrochemical cell (Battery).

Choke Coil: An inductor used to limit the flow of AC.

Circuit: A complete path through which electricity may flow.

Circuit Breaker: A device used to open and close a circuit by automatic means when a predetermined level of current flows through it.

Circular Mil (cm): A measurement of the cross-sectional area of a conductor. kcmil equals 1000 circular mils.

Closed Circuit: A continuous path providing for electrical flow.

Coil: A winding of insulated conductors arranged to produce magnetic flux.

GLOSSARY *(cont.)*

Common Noise: Noise produced between the ground and the hot or the ground and the neutral lines.

Commutator: A series of copper segments connected to the armature of a DC motor.

Conductance: The measure of the ability of a component to conduct electricity expressed in mohs.

Conductor: A substance which offers little resistance to the flow of electrical currents. Insulated copper wire is the most common.

Conduit Body: The part of a conduit system, at the junction of two or more sections of the system, that allows access through a removable cover. Most commonly known as condulets, LBs, LLs, and LRs.

Contacts: The conducting part of a switch that operates with another conducting part to make or break a circuit.

Contactor: A control device that uses a small current to energize or de-energize the load connected to it.

Continuous Load: A load whose maximum current continues for three hours or more.

Core: The central, light-carrying part of an optical fiber.

Coulomb: An electrical current of 1 ampere per second.

Crosstalk: The unwanted energy transferred from one circuit or wire to another which interferes with the desired signal. Usually caused by excessive inductance in a circuit.

Current: The flow of electricity in a circuit, measured in amperes.

Cut-out Box: A surface mounted electrical enclosure with a hinged door.

Cutback Method: A technique for measuring the loss of bare fiber by measuring the optical power transmitted through a long length then cutting back to the source and measuring the initial coupled power.

Cycle: Measured in hertz (Hz), it is the flow of AC in one direction and then in the opposite direction in one time interval completing one wavelength. Hertz measures cycles per second.

GLOSSARY *(cont.)*

Daisy Chaining: The connection of multiple devices in a serial fashion.

Data Rate: The number of bits of information in a transmission system, expressed in bits per second (bps), and which may or may not be equal to the signal or Baud rate.

dB: Decibel referenced to a microwatt.

dBm: Decibel referenced to a milliwatt.

Decibel: 1) A standard logarithmic unit for the ratio of two powers, voltages, or currents. In fiber optics, the ratio is power. 2) Unit for measuring relative strength of a signal parameter such as power or voltage.

Deep Cycle: Battery type that can be discharged to a large fraction of capacity. See Depth of Discharge.

Degauss: To remove residual permanent magnetism.

Delta Connection: A connection that has each coil end connected end-to-end to form a closed loop.

Depth of Discharge (DOD): The percent of the rated battery capacity that has been withdrawn.

Device (Also used as wiring device): The part of an electrical system that is designed to carry, but not use, electrical energy.

Diac: A thyristor that triggers in either direction when its breaker voltage is exceeded.

Diode: Electronic component that allows current flow in one direction only.

Direct Current (DC): Electrical current which flows in one direction only.

DC Compound Motor: The field is connected in both series and shunt with the armature.

DC Permanent Magnet Motor: Uses magnets, not a coil, for the field winding.

DC Series Motor: The field is connected in series with the armature.

DC Shunt Motor: The field is connected in parallel with the armature.

GLOSSARY *(cont.)*

DC Voltage: Voltage that flows in one direction only.

Double Break Contacts: Contacts that break the current in two separate places.

Dispersion: Causes a spreading of light as it propagates through an optical fiber. Three types are modal, material, and waveguide.

Dual Voltage Motor: A motor that operates at more than one voltage level.

Eddy Current: Unwanted current induced in the metal field structure of a motor.

Effective Current: The value of AC which causes the same heating effect as a given value of DC. For sine wave AC, the effective current is 0.7071.

Effective Value: Also called the Root-Mean-Square (RMS), it produces the same I^2R power as an equal DC value.

Efficiency (EFF): The ratio of output power to input power, usually expressed as a percentage.

Electricity: The movement of electrons through a conductor.

Electromagnet: Coil of wire that exhibits magnetic properties when current passes through it.

Electron: The subatomic unit of negative electricity expressed as a charge of 1.6×10^{-19} coulomb.

Electronics: The science of treating charge flow in vacuums, gases and crystal lattices.

Energy: The capacity to do work.

Equalization: The process of restoring all cells in a battery to an equal state of charge.

Ethernet: A 10-Mbps, coaxial standard for LANs. All nodes connect to the cable where they contend for access via CSMA/CD.

Excess Loss: In a fiber-optic coupler, the optical loss from that portion of light that does not emerge from the nominally operational ports.

GLOSSARY *(cont.)*

Excitation: The power required to energize the magnetic field of transformers, generators and motors.

Extrinsic Loss: In a fiber interconnection, that portion of loss that is not intrinsic to the fiber but is related to imperfect joining, which may be caused by the connector or splice.

Farad (F): The unit of measurement of capacitance.

Fault Current: Any current that travels an unwanted path, other than the normal operating path of an electrical system.

Feeder: Circuit conductors between the service and the final branch circuit overcurrent device.

Ferrule: A precision tube that holds a fiber for alignment for interconnection or termination.

Fiber Optics: A technology that uses light as a digital information carrier.

Field: The stationary windings (magnets) of a DC motor.

Filament: A conductor that has a high enough resistance to cause heat.

Filter: A combination of circuit elements which is specifically designed to pass certain frequencies and resist all others.

Flashover: A disruptive electrical discharge around or over (but not through) an insulator.

Fluorescence: The emission of light by a substance when exposed to radiation or the impact of particles.

Flux: An electrical filled energy distributed in space and represented diagrammatically by means of flux lines denoting magnetic or electrical forces.

Footcandle (fc): The amount of light produced by a lamp measured in lumens divided by the area that is illuminated.

Four Wire Circuits: Telephone circuits which use two separate one-way transmission paths of two wires each, as opposed to regular local lines which usually only have two wires to carry conversations in both directions.

GLOSSARY *(cont.)*

Frequency: The number of times per second a signal regenerates itself at a peak amplitude. It can be expressed in hertz (Hz), kilohertz (kHz), megahertz (MHz), etc.

Full-Load Current (FLC): The current required by a motor to produce the full-load torque at the motor's rated speed.

Full-Load Torque (FLT): The torque required to produce the rated power of the motor at full speed.

Fuse: A protective device, also called an OCPD, with a fusible element that opens the circuit by melting when subjected to excessive current.

Gain: A ratio of the amplitude of the output signal to the input signal.

Gap Loss: Loss resulting from the end separation of two axially aligned fibers.

Gb: Gigabit. One billion bits of information.

Gbyte: Gigabyte. One billion bytes of data.

Ghost Voltage: A voltage that appears on a motor that is not connected.

Grid: Term used to describe an electrical utility distribution network.

Ground: An electrical connect (on purpose or accidental) between an item of equipment and the earth.

Grounding: The connection of all exposed non-current carrying metal parts to the earth.

Ground Fault: A condition in which current from a hot power line is flowing to the ground.

Guy: A wire having one end secured and the other fastened to a pole or structure under tension.

Ground Fault Circuit Interrupter (GFCI): An electrical device which protects personnel by detecting hazardous ground faults and quickly disconnects power from the circuit.

Harmonic: 1) A sinusoid which has a frequency which is an integral multiple of a certain frequency. 2) The full multiple of a base frequency.

GLOSSARY *(cont.)*

Heater: A device that is placed in a motor starter to measure the amount of current in the power line.

Heat Sink: A piece of metal used to dissipate the heat of solid-state components mounted on it.

Heating Element: A conductor (wire) that offers enough resistance to produce heat when connected to power.

Henry (H): The unit of measure of inductance in which a current changing its rate of flow one ampere per second induces an electromotive force of one volt.

Hertz (Hz): The unit of measure of the frequency of the AC sine wave to complete a cycle. One Hertz is equal to one cycle of the AC sine-wave per second.

HID Lamp: High Intensity Discharge Lamp.

Holding Current: The minimum current necessary for an SRC to continue conducting.

Horsepower (HP): A unit of power equal to 746 watts that describes the output of electric motors.

Hub: A device which connects to several other devices usually in a star topology. Also called: concentrator, multiport repeater or multi-station access unit (MAU).

Hybrid: An electronic circuit that uses different cable types to complete the circuit between systems.

Impedance (Z): Is the total opposition offered to the flow of AC from any combination of resistance, inductive reactance and capacitive reactance and is measured in Ohms (Ω).

Inductance (L): The property of a circuit that determines how much voltage will be induced into it by a change in current of another circuit and is measured in henrys (H).

Inductive Circuit: A circuit in which current lags voltage.

Inductive Reactance (X_L): The opposition to the flow of AC in a circuit due to inductance and is measured in Ohms (Ω).

GLOSSARY *(cont.)*

In-Phase: The state when voltage and current reach their maximum amplitude and zero level at the same time in a cycle.

Insertion Loss: The loss of power that results from inserting a component, such as a connector or splice, into a previously continuous path.

Insulator: Material that current cannot flow through easily.

Integrated Circuit or IC: A circuit in which devices such as transistors, capacitors, and resistors are made from a single piece of material and connected to form a circuit.

Interface: The point that two systems, with different characteristics, connect.

Isolated Grounded Receptacle: Minimizes electrical noise by providing a separate grounding path.

Isolation Transformer: A one to one transformer that is used to isolate the equipment at the secondary from earth ground.

Jack: A receptacle (female) used with a plug (male) to make a connection to in-wall communications cabling or to a patch panel.

Jacket: The protective and insulating outer housing on a cable. Also called a sheath.

Jogging: The frequent starting and stopping of a motor.

Joule: A unit of electrical energy also called a watt-second; the transfer of one watt for one second.

Jumper: Patch cable or wire used to establish a circuit, often temporarily, for testing or diagnostics.

Junction Box: A box, usually metal, that encloses cable connections.

KB: Kilobyte. One thousand bytes.

Kb: Kilobit. One thousand bits.

Kbps: Kilobits per second. Thousand bits per second.

kcmil: One thousand circular mils.

kVA: Kilovolt-amperes (1,000 volt amps).

GLOSSARY *(cont.)*

kVAR: Kilovar (1,000 reactive volt amps).

kV: Kilovolt (1,000 volts).

kW: Kilowatt (1,000 watts).

kWH: Kilowatt hour**:** The basic unit of electrical energy for utilities equal to a thousand watts of power supplied for one hour.

Lamp: A light source. Reference is to a light bulb, rather than a lamp.

Leakage Current: Current that flows through insulation.

Leg (circuit): One of the conductors in a supply circuit in which the maximum voltage is maintained.

Load: The amount of electric power used by any electrical unit or appliance at any given moment.

Location, Damp: Partially protected locations, such as under canopies, roofed open porches, etc. Also, interior locations that are subject only to moderate degrees of moisture, such as basements, barns, etc.

Location, Wet: Locations underground, in concrete slabs, where saturation occurs, or outdoors.

Locked Rotor: Condition when a motor is loaded so heavily that the shaft cannot turn.

Locked Rotor Current (LRC): The steady-state current taken from the power line with the rotor locked and the voltage applied.

Locked Rotor Torque (LRT): The torque a motor produces when the rotor is stationary and full power is applied.

Loss Budget: The amount of power lost in a fiber optic link. Used in terms of the maximum amount of loss that can be tolerated by a given link.

Loss, optical: The amount of optical power lost as light is transmitted through fiber, splices, couplers, and the like.

Lumen (lm): The unit used to measure the total amount of light produced by a light source.

GLOSSARY *(cont.)*

Mbps: Million bits per second.

Mbyte: Megabyte. Million bytes of information.

Magnetic Field: The invisible field produced by a current-carrying conductor or coil, permanent magnet, or the earth itself, that develops a north and south polarity.

Magnetic Flux: The invisible lines of force that make up the magnetic field.

Magnetic Induction: The setting up of magnetic flux lines in a material by an electric current. The number of lines is measured in maxwells.

Maxwell: The unit of measurement of the total number of magnetic flux lines in a magnetic field.

Mechanical Splice: A semipermanent connection between two fibers made with an alignment device and index matching fluid or adhesive.

Micron (m): A unit of measure, 10^{-6} m, used to measure wavelength of light.

Mode: A single electromagnetic field pattern that travels in fiber.

Motor: A machine that develops torque (rotating mechanical force) on a shaft to produce work.

Motor Efficiency: The effectiveness of a motor to convert electrical energy into mechanical energy.

Motor Starter: An electrically operated switch (contactor) that includes overload protection.

Motor Torque: The force that produces rotation in the shaft.

Multiplex: To combine multiple input signals into one for transmission over a single high-speed channel. Two methods are used: (1) frequency division, and (2) time division.

Mutual Inductance: The effect of one coil inducing a voltage into another coil.

Nanometer (nm): A unit of measure, 10^{-9} m, used to measure the wavelength of light.

GLOSSARY *(cont.)*

NEC®: National Electrical Code®, which contains safety rules for all types of electrical installations.

No-Load Current: The current demand of a transformer primary when no current demand is made on the secondary.

Normally Open Contacts: Contacts that are open before being energized.

Normally Closed Contacts: Contacts that are closed before being energized.

Ohm: The unit of measurement of electrical resistance. One ohm of resistance will allow one ampere of current to flow through a pressure of one volt.

Ohm's Law: A law which describes the mathematical relationship between voltage, current and resistance.

Open Circuit: A condition that provides no path for electric current to flow in a circuit.

Open Circuit Voltage: The maximum voltage produced by a photovoltaic cell, module, or array without a load applied.

Optical Power: The amount of radiant energy per unit of time, expressed in linear units of watts or on a logarithmic scale, in dBm (where 0 dB = 1 mW) or dB (where 0 dB = 1 W).

Oscillation: Fluctuations in a circuit.

Out of Phase: Having AC sine waveforms that are of the same frequency, but are not passing through corresponding values at the same instant.

Outlet: The place in the wiring system where the current is taken to supply equipment.

Overcurrent: Too much current.

Overload Protection: A device that prevents overloading a circuit or motor such as a fuse or circuit breaker.

Parallel Circuit: Contains two or more loads and has more than one path through which current flows.

GLOSSARY *(cont.)*

Peak Value: The maximum value of either the positive or negative alteration of a sine wave.

Period: The time required to produce one cycle of a waveform.

Phase: The fractional part of a period through which the time variable of a periodic quantity has moved.

Phase Converter: A device that derives three phase power from single phase power.

Phase Shift: The state when voltage and current in a circuit do not reach their maximum amplitude and zero level at the same time.

Photovoltaic: Changing light into electricity.

Pigtail: A short length of fiber permanently attached to a component.

Polarity: The particular state of an object, either positive or negative, which refers to the two electrical poles, north, and south.

Power (Watts): A basic unit of electrical energy, measured in watts.

Power Budget: The difference (in dB) between the transmitted optical power (in dBm) and the receiver sensitivity (in dBm).

Power Factor: The ratio of true power (kW) to apparent power (kVA) for any given load and period of time.

Primary Winding: The coil of a transformer which is energized from a source of alternating voltage and current. The input side.

Pulse Spreading: The dispersion of an optical signal with time as it propagates through an optical fiber.

Reactance (X): The measure of the induced voltage due to the inductance of a circuit and is measured in Ohms (Ω).

Relay: A device that controls one electrical circuit by opening and closing the contacts in another circuit.

Resistance (R): The opposition to the flow of current in an electrical circuit and is measured in Ohms (Ω).

GLOSSARY *(cont.)*

Resistive Circuit: A circuit containing resistive loads such as heating elements or incandescent lamps.

Resonance (f): A condition in a circuit where the frequency of an externally applied force equals the natural tendency of the circuit; when the inductive reactance (X_L) equals capacitive reactance (X_C). Resonance is measured in Hertz (Hz).

RG-58: The coaxial cable used by Thin Ethernet (10Base2). It has a 50 ohm impedance and so must use 50 ohm terminators.

RG-59: The coaxial cable, with 75 ohm impedance, used in cable TV and other video environments.

RG-62: The coaxial cable, with 93 ohm impedance, used by ARCNet and IBM 3270 terminal environments.

RJ11: A standard six conductor modular jack or plug that uses two to six conductors.

RJ22: The standard four conductor modular jack that connects a telephone handset to its base unit.

RJ45: A standard eight conductor modular jack or plug that uses two to eight conductors.

Root-Mean-Square: A mathematical expression equal to 0.707 times the peak value of an AC waveform.

Rotary Convertor: A type of phase convertor.

Rotor: The rotating part of an AC motor.

Secondary Winding: The winding of a transformer that receives electrical energy by electromagnetic induction from the primary winding and then delivers it to a load. The output side of a transformer.

Series Circuit: A circuit that has only one current path to and from the power source.

Short Circuit: An undesired path for electrical current.

Shunt: Denotes a parallel connection.

GLOSSARY *(cont.)*

Sine Wave: A waveform corresponding to a single-frequency, periodic oscillation, which can be shown as a function of amplitude against angle and in which the value of the curve at any point is a function of the sine of that angle.

Single-Phase Power: One of the three alternating currents in a circuit.

Solenoid: An electromagnet with a movable iron core.

Split-Phase Motor: A single phase AC motor that has a running and a starting winding.

Split-Wired Receptacle: A receptacle that has the metal tap removed between the hot terminals.

Stator: The stationary part of an AC motor.

Switch (electrical): A device to start or stop the flow of electricity.

Surge Capacity: The requirement of an invertor to tolerate a momentary current surge imposed by starting AC motors or transformers.

T1: A digital carrier facility for transmitting a single DS1 digital stream over two pairs of regular copper telephone wires at 1.544 Mbps.

T2: A digital carrier facility used to transmit a DS2 digital stream at 6.312 Mbps.

T3: A digital carrier facility used to transmit a DS3 digital stream at 44.746 megabits per second.

T4: A digital carrier facility used to transmit a DS4 digital stream at 273m bps.

Taps: Connecting points on a transformer coil.

Thermal Protection: Refers to an electrical device which has inherent protection from overheating. Typically in the form of a bimetal strip which bends when heated to a certain point.

Three-Phase Power: A combination of three alternating currents (usually denoted as a, b and c) in a circuit with their voltages displaced 120° or one-third of a cycle.

GLOSSARY *(cont.)*

Transformer: A device which uses magnetic force to transfer electrical energy from one coil of wire to another. In the process, transformers can also change the voltage at which this electrical energy is transmitted.

True Power (P_T): The actual power used in an electrical circuit measured in watts (w) or kilowatts (kW).

Turns Ratio: The ratio between the voltage and the number of turns on the primary and secondary windings of a transformer.

VA: Volt-amperes.

Volt (E) or (V): The unit of measurement of electrical pressure (force). One volt will force one ampere of current to flow through a resistance of one ohm.

Voltage Drop: Voltage reduction due to wire resistance.

Watt (W): The unit of measurement of electrical power produced by a current of one amp across a potential difference of one volt.

Waveform: Characteristic shape of an electrical current or signal.

Wavelength: The distance between the same two points on adjacent waves; the time required for a wave to complete a single cycle.

Wye Connection: A connection that has one end of each coil connected together and the other end of each coil left open for external connections.

About The Author

Paul Rosenberg has an extensive background in the electrical, data, construction and fiber optic trades. He is a leading voice in the electrical industry with years of experience from apprentice to project manager. Paul has written for all of the leading electrical and low voltage industry magazines and has authored more than 50 books.

In addition, he wrote the first standard for the installation of optical cables (ANSI-NEIS-301) and was awarded a patent for a power transmission module. Paul taught nineteen courses for Iowa State University and works as a consultant and expert witness in legal cases. He speaks occasionally at industry events.